Contributions to Antarctic Research II

Physical Sciences

ANTARCTIC OCEANOLOGY
Joseph L. Reid, *Editor*
ANTARCTIC OCEANOLOGY II: THE AUSTRALIAN-NEW ZEALAND SECTOR
Dennis E. Hayes, *Editor*

ANTARCTIC SNOW AND ICE STUDIES
Malcolm Mellor, *Editor*
ANTARCTIC SNOW AND ICE STUDIES II
A. P. Crary, *Editor*

ANTARCTIC SOILS AND SOIL FORMING PROCESSES
J. C. F. Tedrow, *Editor*
DRY VALLEY DRILLING PROJECT
L. D. McGinnis, *Editor*
GEOLOGICAL INVESTIGATIONS IN NORTHERN VICTORIA LAND
Edmund Stump, *Editor*
GEOLOGY AND PALEONTOLOGY OF THE ANTARCTIC
Jarvis B. Hadley, *Editor*
GEOLOGY OF THE CENTRAL TRANSANTARCTIC MOUNTAINS
Mort D. Turner and John F. Splettstoesser, *Editors*
GEOMAGNETISM AND AERONOMY
A. H. Waynick, *Editor*
METEOROLOGICAL STUDIES AT PLATEAU STATION, ANTARCTICA
Joost A. Businger, *Editor*
OCEANOLOGY OF THE ANTARCTIC CONTINENTAL SHELF
Stanley S. Jacobs, *Editor*
STUDIES IN ANTARCTIC METEOROLOGY
Morton J. Rubin, *Editor*
UPPER ATMOSPHERE RESEARCH IN ANTARCTICA
L. J. Lanzerotti and C.G. Park, *Editors*
THE ROSS ICE SHELF: GLACIOLOGY AND GEOPHYSICS
C. R. Bentley and D. E. Hayes, *Editors*
VOLCANOES OF THE ANTARCTIC PLATE AND SOUTHERN OCEANS
W. E. LeMasurier and J. T. Thomson, *Editors*
MINERAL RESOURCES POTENTIAL OF ANTARCTICA
John F. Splettstoesser and Gisela A. M. Dreschhoff, *Editors*
CONTRIBUTIONS TO ANTARCTIC RESEARCH I
David H. Elliot, *Editor*

American Geophysical Union

ANTARCTIC RESEARCH SERIES

Biological and Life Sciences

BIOLOGY OF THE ANTARCTIC SEAS
Milton O. Lee, *Editor*
BIOLOGY OF THE ANTARCTIC SEAS II
George A. Llano, *Editor*
BIOLOGY OF THE ANTARCTIC SEAS III
George A. Llano and Waldo L. Schmitt, *Editors*
BIOLOGY OF THE ANTARCTIC SEAS IV
George A. Llano and I. Eugene Wallen, *Editors*
BIOLOGY OF THE ANTARCTIC SEAS V
David L. Pawson, *Editor*
BIOLOGY OF THE ANTARCTIC SEAS VI
David L. Pawson, *Editor*
BIOLOGY OF THE ANTARCTIC SEAS VII
David L. Pawson, *Editor*
BIOLOGY OF THE ANTARCTIC SEAS VIII
David L. Pawson and Louis S. Kornicker, *Editors*
BIOLOGY OF THE ANTARCTIC SEAS IX
Louis S. Kornicker, *Editor*
BIOLOGY OF THE ANTARCTIC SEAS X
Louis S. Kornicker, *Editor*
BIOLOGY OF THE ANTARCTIC SEAS XI
Louis S. Kornicker, *Editor*
BIOLOGY OF THE ANTARCTIC SEAS XII
David L. Pawson, *Editor*
BIOLOGY OF THE ANTARCTIC SEAS XIII
Louis S. Kornicker, *Editor*
BIOLOGY OF THE ANTARCTIC SEAS XIV
Louis S. Kornicker, *Editor*
BIOLOGY OF THE ANTARCTIC SEAS XV
Louis S. Kornicker, *Editor*
BIOLOGY OF THE ANTARCTIC SEAS XVI
Louis S. Kornicker, *Editor*
BIOLOGY OF THE ANTARCTIC SEAS XVII
Louis S. Kornicker, *Editor*
BIOLOGY OF THE ANTARCTIC SEAS XVIII
Louis S. Kornicker, *Editor*
BIOLOGY OF THE ANTARCTIC SEAS XIX
Louis S. Kornicker, *Editor*
BIOLOGY OF THE ANTARCTIC SEAS XX
Louis S. Kornicker, *Editor*
BIOLOGY OF THE ANTARCTIC SEAS XXI
Louis S. Kornicker, *Editor*

ANTARCTIC TERRESTRIAL BIOLOGY
George A. Llano, *Editor*
TERRESTRIAL BIOLOGY II
Bruce Parker, *Editor*
TERRESTRIAL BIOLOGY III
Bruce Parker, *Editor*

ANTARCTIC ASCIDIACEA
Patricia Kott
ANTARCTIC BIRD STUDIES
Oliver L. Austin, Jr., *Editor*
ANTARCTIC PINNIPEDIA
William Henry Burt, *Editor*
ANTARCTIC CIRRIPEDIA
William A. Newman and Arnold Ross
BIRDS OF THE ANTARCTIC AND SUB-ANTARCTIC
George E. Watson
ENTOMOLOGY OF ANTARCTICA
J. Linsley Gressitt, *Editor*
HUMAN ADAPTABILITY TO ANTARCTIC CONDITIONS
E. K. Eric Gunderson, *Editor*
POLYCHAETA ERRANTIA OF ANTARCTICA
Olga Hartman
POLYCHAETA MYZOSTOMIDAE AND SEDENTIARIA OF ANTARCTICA
Olga Hartman
RECENT ANTARCTIC AND SUBANTARCTIC BRACHIOPODS
Merrill W. Foster

Volume 53 | ANTARCTIC RESEARCH SERIES

Contributions to Antarctic Research II

David H. Elliot, Editor

American Geophysical Union
Washington, D.C.
1991

Volume 53 | **ANTARCTIC RESEARCH SERIES**

CONTRIBUTIONS TO ANTARCTIC RESEARCH II

David H. Elliot, Editor

Library of Congress Cataloging in Publication Data

(REvised for volume 2)

Contributions to Antarctic research.
(Antarctic research series, ISSN 0066-4634 ; v. 53)
Vol. 2 edited by David H. Elliott.
ISBN 0-87590-760-1 (v. 1)
ISBN 0-87590-809-8 (v. 2)
1. Antarctic regions. I. Elliot, David H. II. American Geophysical Union.
III. Series: Antarctic research series ; v. 53, etc.
G860.C58 1990 90-1279
998′.9 CIP

ISSN 0066-4634

Published by
American Geophysical Union
With the aid of grant DPP-89-15494 from the
National Science Foundation

Printed in the United States of America

CONTENTS

The Antarctic Research Series:
STATEMENT OF OBJECTIVES

The Antarctic Research Series provides for the presentation of detailed scientific research results from Antarctica, particularly the results of the United States Antarctic Research Program, including monographs and long manuscripts.

The series is designed to make the results of Antarctic fieldwork available. The Antarctic Research Series encourages the collection of papers on specific geographic areas within Antarctica. In addition, many volumes focus on particular disciplines, including marine biology, oceanology, meteorology, upper atmosphere physics, terrestrial biology, geology, glaciology, human adaptability, engineering, and environmental protection.

Topical volumes in the series normally are devoted to papers in one or two disciplines. Multidisciplinary volumes, initiated in 1990 to enable more rapid publication, are open to papers from any discipline. The series can accommodate long manuscripts and utilize special formats, such as maps.

Priorities for publication are set by the Board of Associate Editors. Preference is given to research manuscripts from projects funded by U.S. agencies. Because the series serves to emphasize the U.S. Antarctic Research Program, it also performs a function similar to expedition reports of many other countries with national Antarctic research programs.

The standards of scientific excellence expected for the series are maintained by the review criteria established for the AGU publications program. Each paper is critically reviewed by two or more expert referees. A member of the Board of Associate Editors may serve as editor of a volume, or another person may be appointed. The Board works with the individual editors of each volume and with the AGU staff to assure that the objectives of the series are met, that the best possible papers are presented, and that publication is timely.

Proposals for volumes or papers offered should be sent to the Board of Associate Editors, Antarctic Research Series, at 2000 Florida Avenue, N.W., Washington, D.C. 20009. Publication of the series is partially supported by a grant from the National Science Foundation.

Board of Associate Editors
Antarctic Research Series

Revised 5/13/91

CONTRIBUTIONS TO ANTARCTIC RESEARCH II
ANTARCTIC RESEARCH SERIES, VOLUME 53, PAGES 1–17

PEDOGENIC LINKAGE BETWEEN THE COLD DESERTS OF ANTARCTICA AND THE POLAR DESERTS OF THE HIGH ARCTIC

J. C. F. TEDROW

Department of Environmental Resources, Rutgers University, New Brunswick, New Jersey 08903

The Cold Deserts of Antarctica and the Polar Deserts of the High Arctic, the latter of which includes the Queen Elizabeth Islands, northern Greenland, Svalbard, Franz Josef Land, Severnaya Zemlya, and a few other localities, have cryogenic soils with desertlike features. Both sectors have soils with a mineral appearance, desert pavement, small to nonexistent organic components, and low ground temperatures. Both are underlain by permafrost and are subject to cryogenic processes. Whereas virtually all of the soils in Antarctica are well drained, in the High Arctic the soil pattern consists of a mosaic of Polar Desert soils, as well as various hydric varieties including shallow bogs. In the Polar Deserts the landforms, especially the surficial deposits, are much younger chronologically than their Cold Desert counterparts. Polar Desert soils generally have a more acid reaction than Cold Desert soils. Moisture regimes are quite different in the two sectors. Cold Desert soils have a dry condition even down to the frost table. Polar Desert soils are moist during early summer thaw, however, and are subjected to occasional summer rainfall and probably receive some moisture from condensation; the zone above the receding frost table is usually quite wet. The humus component in Cold Desert soils is usually nil, but in the Polar Desert soils it is commonly as much as 1% to 2% or higher, especially in medium-textured soils. Whether there are enough distinct differences between the two soils to recognize two different categories remains somewhat moot. After equating all information, however, one appears to be justified in tentatively recognizing two distinct soils.

INTRODUCTION

The north and the south polar regions have a number of characteristics in common, particularly extremely low seasonal temperatures, the presence of perennially frozen ground, small quantities of precipitation, existing glaciers, and related cryogenic features. In the far north (High Arctic) the vascular plant cover exists mainly as clumps and patches with the remaining space having a barren, mineral appearance. In Antarctica conditions are noticeably colder and drier, however, and, accordingly, one may travel the ice-free areas for miles without any trace of vascular plants being evident.

A question that naturally arises is the relation of the soil-forming processes of the High Arctic to those of Antarctica. Putting it another way, are the soils and soil-forming processes within the two regions mirror images of each other? This question was addressed by *Markov* [1956], *McCraw* [1960], *Tedrow* [1968, 1977], *Everett* [1976], *Ugolini* [1986*a*], *Bockheim* [1979, 1980], *Campbell and Claridge* [1987], and others. This report reviews the subject of soil development in the High Arctic and in Antarctica and makes a comparative analysis as to processes and resultant morphologies of soils in the two regions. The term "High Arctic" is used in this report in a nonspecific sense. But, in general, "High Arctic" refers to land areas north, about 76°–78°N, and includes the Queen Elizabeth Islands, northern Greenland, Svalbard, Franz Josef Land, Servernaya Zemlya, and a few other islands of the Soviet Arctic archipelago. The above locations generally exist within the 4°C July isotherm. Earlier, *Nordenskjöld and Mecking* [1928] used the 5°C July isotherm for delineating the High Arctic.

ANTARCTIC SOIL PROCESSES AND CLASSIFICATION

Estimates generally indicate that less than 3% of Antarctica is ice-free, with most of such areas consisting of only exposed bedrock and boulders (Figure 1). In using the term "soil" in Antarctica I am referring to less than 1% of the total area of the continent having any form of genetic soil.

Jensen [1916] is credited with the first report on Antarctic soils in that he made chemical analyses of samples from Cape Royds on Ross Island. He stated that there was no soil in the traditional sense, but in the soillike material there was evidence of mineral alteration and an abundant supply of calcium, potassium, and phosphorus, much of which was related to the presence of local alkali trachyte and salt spray. The work of *Markov* [1956] in the Mirnyy area of Queen Mary Coast is of special significance because he stated that a migration of salts occurs

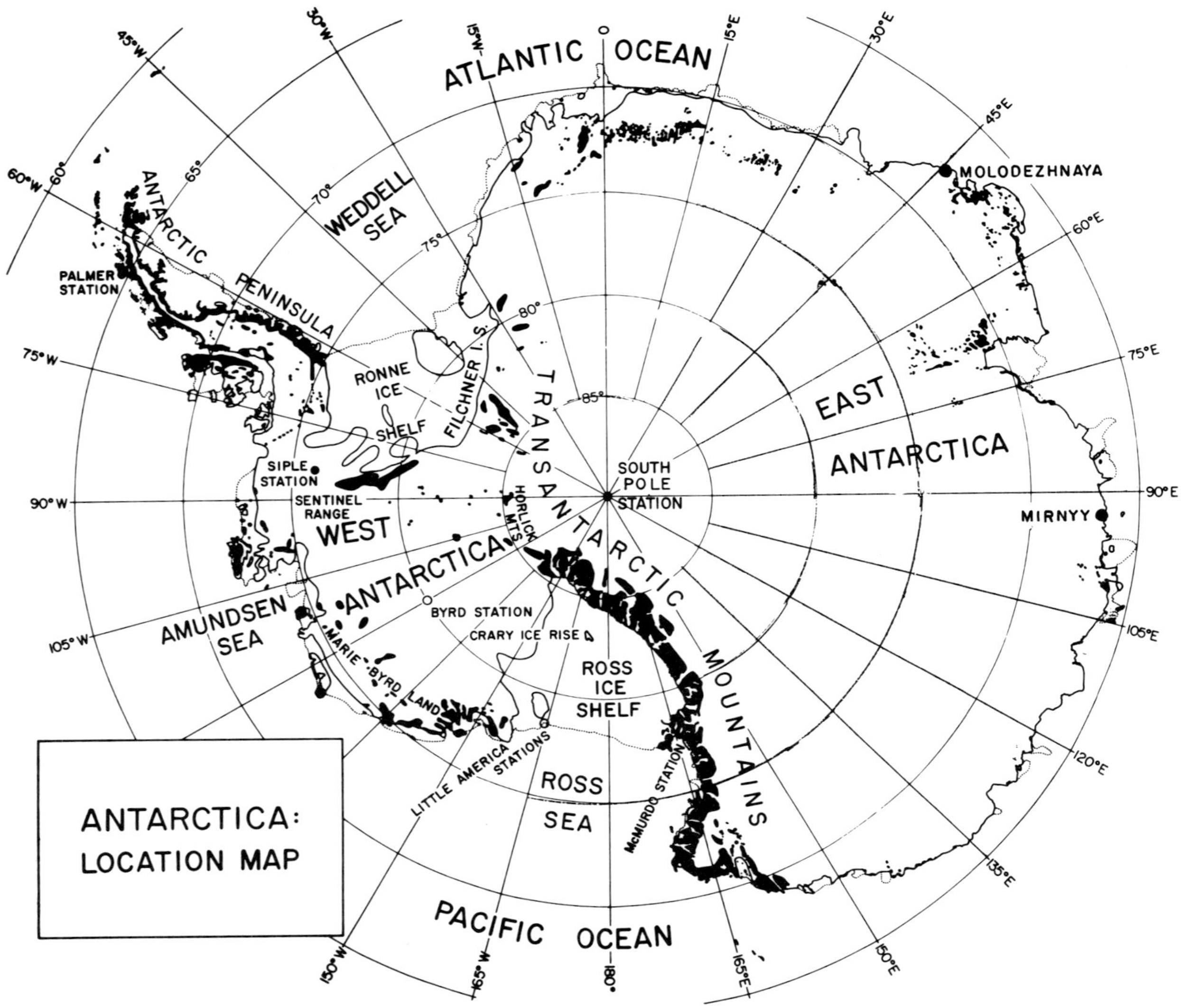

Fig. 1. Map of Antarctica. The shaded areas are largely ice-free. (Courtesy of C. R. Bentley.)

along temperature gradients within the rocks, which in turn releases soluble salts for redistribution in the soils. Apparently, Markov was the first to use the term "Cold Desert" (factually his term was "Antarctic Cold Desert"). *Glazovskaia* [1958] continued investigations relating to weathering and primary soil formation in the Mirnyy sector of Antarctica and, in addition, provided much evidence of physical and chemical weathering, soluble salt production, desert varnish formation, and the formation of weathering crusts on rock surfaces. There was also convincing evidence of primary mineral alteration, such as feldspar weathering, hornblendes turning brownish, chlorites becoming discolored and covered by ferruginous films, and garnets splitting and the cracks becoming filled with an iron-stained substance. Another aspect of Glazovskaia's work deals with biotic elements, in that she reported the presence of green films of unicellular algae, not only in the cracks of exfoliated rock surfaces but also in the depths of the rocks. The work of *Avsiuk et al.* [1956] largely mirrored the Cold Desert concept as proposed in the report by Markov.

J. McCraw's work in south Victoria Land marks the initial step in providing for soil systematics in Antarctica. He displayed a soil map of Taylor Valley at the Seventh International Soil Science Congress in Madison, Wisconsin, in 1960, and later published the map and a report showing soils at the equivalent of the series level [*McCraw*, 1967]. Earlier, *McCraw* [1960] addressed the problem of a soil grouping based on genetic principles (Table 1). His work recognized carbonate accumulation, soluble salt concentrations, weathering, desert varnish development, the formation of soil structure, and a desert process.

Continuing with his work mainly in south Victoria Land, *McCraw* [1967] proposed that the Zonal soils be termed "frigic," a term proposed earlier by *Claridge* [1965]; and, accordingly, McCraw suggested that the higher taxonomic units be recognized as follows:

Frigic (Zonal) Soils

1. With accumulations of calcium carbonate and gypsum at or near the surface and the presence of soluble salts.

2. With calcium carbonate, gypsum, and soluble salts, fairly evenly distributed.

Intrazonal Soils

3. With a functional organic component and those present around fumaroles.

Azonal Soils

4. Steep areas, windblown and other deposits.

About the same time that McCraw was developing his classification system for Antarctic soils, *Tedrow and Ugolini* [1966], conducting investigations in the dry valley system of south Victoria Land, proposed their own generalized classification scheme. Using the concepts of Markov and Glazovskaia, they designated all Antarctic soils as belonging to the Cold Desert zone. Six genetic sets of soils were recognized at the higher taxonomic levels as follows:

Cold Desert soils (Antarctica)

1. Ahumic soils—somewhat analogous to *Nikiforoff*'s [1937] term "abiotic" but with the desert process taking place at low temperatures.

2. Evaporite soils—soils of the closed basins and other low areas.

3. Protoranker soils—initial soil formation with a detectable humus component, the term being from *Kubiëna* [1953].

4. Ornithogenic soils—organic soils of the rookeries in which the organic component is derived from animal droppings, feathers, bones, etc. The term is used in the sense of *Syroechovsky* [1959].

5. Lithosols

6. Regosols

Throughout this report the term "Cold Desert" soil is used in a restricted sense to typify mainly the Ahumic (Frigic) soils of the Cold Desert.

The studies of *Campbell and Claridge* [1969] were especially important in establishing an overall classification scheme for Antarctic soils. Following a series of preliminary reports they proposed the grouping shown in Table 2. Their work not only established the main qualitative soil processes associated with Antarctica as a whole but also recognized some quantitative factors as exemplified by dividing the "Zonal" soils into Ultraxerous, Xerous, and Subxerous varieties.

Pedologic zonation of Antarctica has been rather tenuous. *MacNamara* [1969*a*, *b*], *Campbell and Claridge* [1969], and *Everett* [1976] recognized certain subzonal genetic soil varieties within the continent. *Korotkevich* [1967] distinguished three subzones for Antarctica: (1) a northern subzone which occupies the northeast coast of the Antarctic Peninsula and nearby islands, (2) an intermediate subzone which covers the coastal oasis and islands, and (3) a southern

TABLE 1. Soils of the Ross Dependency, Antarctica

Group	Topography and Parent Material	Characteristics
I	Fans, floodplains, and terraces of recent alluvium; surface smooth with few narrow frost cracks	No secondary carbonate horizons; some salt; no stone pavement; sparse moss; slight physical weathering.
Ia	Intermediate terraces; surface smooth but well-developed raised border polygons	Slightly developed secondary carbonate horizon at surface; slightly developed stone pavement; little physical weathering
	Scree receiving meltwater from above; much solifluction; stone stripes, no polygons	Slightly developed carbonate horizon at surface; no pavement
	Scree receiving no meltwater, mainly large blocks with sparse soils in crevices	Slightly developed carbonate horizon at surface; no pavement
II	Valley floor moraines, including thin moraine over fine lacustrine sediments; high terraces; scoria slopes (Ross Island); surface undulating with well-developed polygons	Distinct secondary carbonate horizon near surface; well-developed stone pavement; platy structure in topsoils and "silt ball" subsoils in fine sediment; much physical weathering
III	Lateral moraines at intermediate levels; polygons rare	Weakly developed carbonate horizon near surface; extensive physical weathering; boulders marked by cavernous weathering; desert varnish on stones; no salt
IV	Moraines at high levels; no polygons; surface smooth, no large boulders, no stone pavement	No secondary carbonate horizon; some salt in lower horizons; very extensive physical weathering—most rocks except dolerite disintegrated; desert varnish common

Source: *McCraw* [1960].

TABLE 2. A Classification of Frigic Soils

Available Moisture Status	Degree of Development	Parent Material	
		Material	Composition
	Zonal Soils (Frigic Soils)		
Ultraxerous	Weakly developed	Alluvium	Greywacke
Xerous	Moderately developed	Colluvium	Schist
Subxerous	Strongly developed	Moraine	Granite
		Massive rock, etc.	Dolerite, etc.
	Intrazonal Soils		
Soils formed under the influence of saline groundwater	Evaporite soils		
	Algal peats		
Soils with a dominant organic constituent	Avian soils (ornithogenic soils)		
Soils formed under the influence of hydrothermal waters	(Hydrothermal soils)		
	Azonal Soils		
Soils of recent beaches, fans, streambeds, etc.	Recent soils		

Source: *Campbell and Claridge* [1969].

subzone which consists of the interior of the continent. *Tedrow* [1977] indicated the probability of a Polar Desert zone existing in the northern Antarctic Peninsula and the remainder of the continent was designated as the Cold Desert zone. *Bockheim and Ugolini* [1990] published a map of Antarctica showing virtually all of the continent as belonging to an Antarctic Cold Desert zone with the coastal sector of East Antarctica extending from Wilkes Land to Molodezhnaya plus the northern part of the Antarctic Peninsula near Graham Land as a probably Antarctic Polar Desert zone. At the extreme northern extension of the Antarctic Peninsula they designated a Subantarctic tundra zone.

Tedrow [1977] proposed a classification system for Antarctic soils (Table 3) that, in principle, was similar to that of *McCraw* [1960], *Tedrow and Ugolini* [1966], and *Campbell and Claridge* [1969].

In connection with Table 3 it is important to mention the work of *MacNamara* [1969*a*, *b*] in Enderby Land, East Antarctica. Instead of soils resembling those of the dry valleys, the Beardmore Glacier area, and the Queen Mary Coast, MacNamara found soils that had brownish, reddish yellow, and olive-yellow colors, which were derived from greater weathering. They also were somewhat leached. Accordingly, MacNamara introduced the terms "Red Ahumisol" and "Brown Ahumisol." The findings of

TABLE 3. A Proposed Classification Scheme for Soils of the Cold Deserts

First Order	Second Order	Third Order	Fourth Order	Fifth Order
Cold Desert Soil Zone	Ahumic (Frigic)† soils Ultraxerous† Xerous† Subxerous† Ahumisol‡ Evaporite soils Ornithogenic (Avian)† Soils§ Other soils Protoranker Algae peats† Hydrothermal soils‖ Regosols (recent soils) Lithosols	Separations based on textural and mineral properties of the parent material, etc.	Not Applicable for the mainland of Antarctica	Soil type + patterned ground

Source: *Tedrow* [1977]. The northern Antarctic Peninsula and the Antarctic islands are likely to be allied to the polar desert, subpolar desert, and tundra zones for purposes of classification.

†*Campbell and Claridge* [1969].

‡*MacNamara* [1969*a*].

§*Syroechkovsky* [1959].

‖*Ugolini and Starkey* [1966].

MacNamara also provide an avenue of possible genetic linkage between soils of Antarctica and those of the High Arctic. *Everett* [1976], in the vicinity of the South Shetland Islands and the northern part of the Antarctic Peninsula, also showed that some of the soils had general features in common with those of the High Arctic.

In addition to the generalized regional pattern of soil development on a climatic basis, there have been a number of important findings in Antarctica relative to the effect of time, topography, material, and site position on soil development. With increased age the soils acquire somewhat brownish or yellowish colors and tend to have larger amounts of carbonates, sulfates, and other soluble salt forms [e.g., *Ugolini*, 1965, 1970, 1986*b*; *Tedrow and Ugolini*, 1966; *Everett*, 1971; *Behling*, 1971; *Bockheim*, 1979; *Pastor and Bockheim*, 1980; *Campbell and Claridge*, 1987; *Bockheim et al.*, 1990]. *Pastor and Bockheim* [1980] described the soils of Taylor Valley, Antarctica, as having a desert pavement, poorly developed horizons followed at depth with permafrost which may be dry or ice cemented. They describe the soils as having an alkaline reaction and being enriched with such salts as mirabilite, tachyhydrite, and halite.

HIGH ARCTIC SOILS: PROCESSES AND CLASSIFICATION

Developments relating to soil classification in the High Arctic were somewhat different from those in Antarctica. In the Antarctic nearly all important developments in soil science postdate 1955, and there have been few differences in professional opinions as to the pedogenic processes taking place. Also, with Antarctica being isolated from other landmasses, it has been treated on a megascopic scale largely as one climatic region. The High Arctic exists as the northernmost sector of a climatic continuum, with noncontinuous landmasses and an indistinct southern boundary. Another factor that is pertinent to High Arctic studies is that investigators came from several nations, particularly the Soviet Union, Germany, and Denmark, followed by the United States, Canada, France, Sweden, and Poland. Lines of communication were somewhat imperfect; none agreed upon a common denominator for presenting pedologic information, a situation that continues up to the present.

As early investigators penetrated the High Arctic, some approached via Siberia and others went directly to Svalbard or Greenland. Landscape elements that were being evaluated were therefore not always entirely equivalent, which added to the complexity of the problem.

Egede [1745] recognized the colder and drier aspects of the High Arctic and since that time such terms as polar desert, arctic desert, rock desert, fell field, cold steppe, polar tundra, barren ground tundra, and others have been used to depict the cold, barren conditions of the High Arctic. The term "Polar Desert" was first introduced by *Passarge* [1920] as a designation for the highest northern latitudes where the temperature of the warmest month does not exceed 2°C. He used the term "desert" because the vegetation in the parts not covered by ice was very poor, and the favorable sites were colonized mainly by lichens and mosses. Some subsequent investigators not only delineated such a zone but, in addition, subdivided it as exemplified by the work of *Ivanova and Rosov* [1962], *Korotkevich* [1966], and *Aleksandrova* [1988]. Others, as depicted by *Walton* [1972], *Bliss* [1977], *Makeev* [1978], and *Ugolini* [1986*a*], appear to be in general agreement as to the use of the term "Polar Desert." *Bovis and Barry* [1974] delineated deserts and semideserts of the far north on climatic criteria. Their designation of Polar Desert included the western Canadian Arctic archipelago, northern Ellesmere Island, northern Greenland, Franz Josef Land, Severnaya Zemlya, and Novo Sibirskie.

To establish a framework for describing soils of the High Arctic—a term used by this writer as nearly equivalent to the area designated as the Polar Desert soil zone—it is necessary to designate where the Polar Desert soil zone occurs. Figure 2 shows the Polar Desert zone as depicted by *Nordenskjöld and Mecking* [1928], *Korotkevich* [1967], *Tedrow* [1977], and *Aleksandrova* [1988].

Unlike conditions in Antarctica, where there are few soils with impeded drainage, the High Arctic is mantled with a mosaic of well-drained soils plus others having some form of wetness, including shallow peat deposits. The latter conditions are brought about by a number of factors such as relatively higher precipitation values (approximately 125–150 mm of water equivalent per year), relatively higher humidity, plus the generation of meltwater and late snow cover that, collectively, provides for a constant early summer source of moisture for lower terrain and some of the sloping lands as well. Further, many of the more recently deglaciated landscapes do not have a well-developed drainage system, a condition that leads to a tendency for formation of poorly drained soils. Figure 3 typifies a cross section of a High Arctic landscape in which the higher position is mantled with a desertlike soil, but the sloping land has hummocky ground, which would probably be considered as a paravariety of Tundra soil. The lower areas are shown as Polar Desert–Tundra Interjacence followed downslope by Tundra and Bog soils with a decided hydric aspect [*Tedrow*, 1970, 1974; *Tedrow et al.*, 1968].

Liverovskij [1934] and *Gorodkov* [1935, 1939] are generally credited with the earliest usage of the term "Polar Desert" pedologically, but there were subsequent questions raised as to whether Liverovskij's work was actually within the Polar Desert zone. Early investigations in Svalbard (Spitsbergen) by *Meinardus* [1912] and by *Blanck* [1919] contained valuable mineral/chemical parameters of the desertlike soils, but their findings as to pedogenic processes were inconclusive and there was little reference to soil

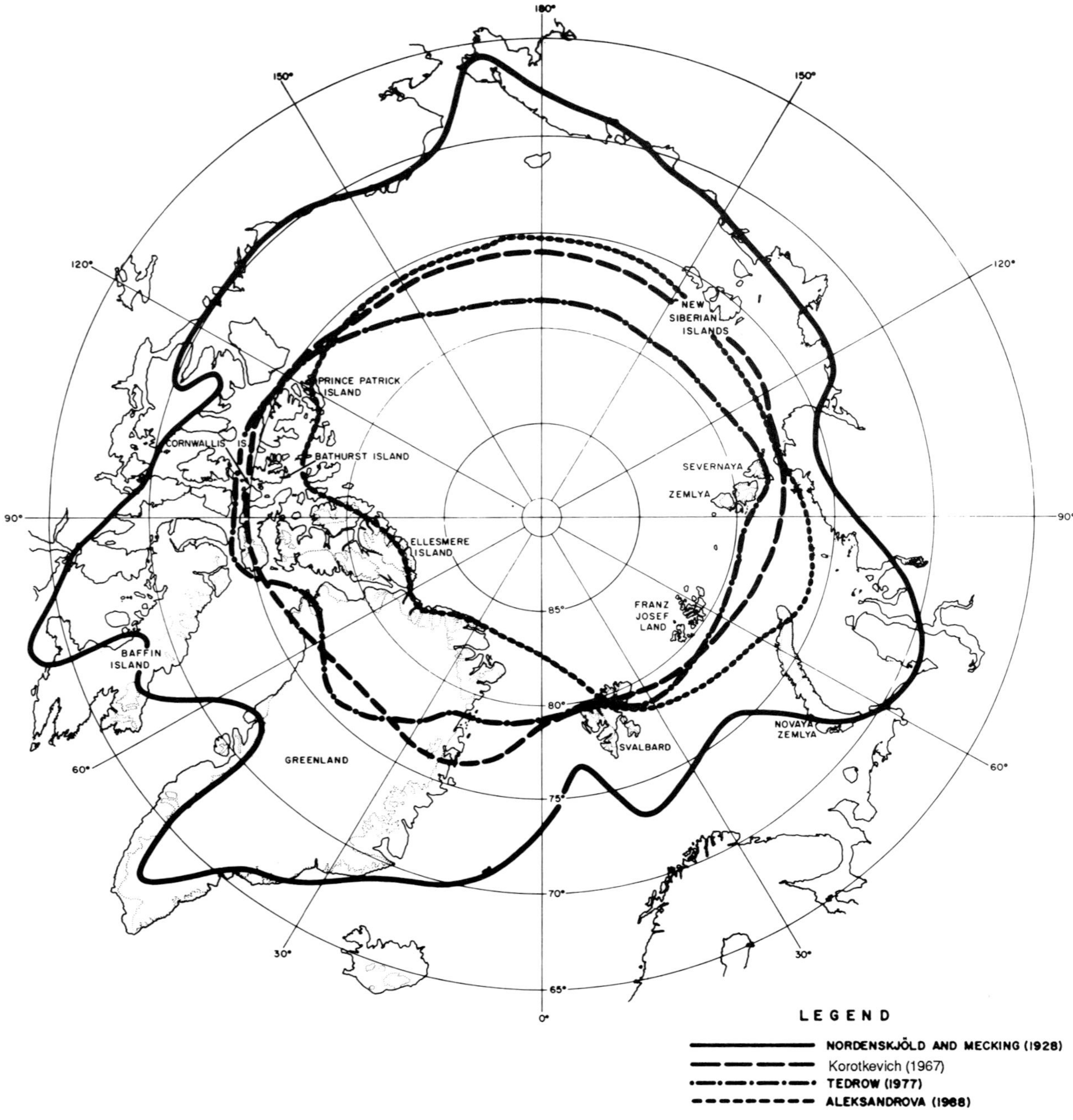

Fig. 2. Map of the northern latitudes, showing the Polar Desert sector as depicted by various authors. (Sources: *Nordenskjöld and Mecking* [1928], *Korotkevich* [1967], *Tedrow* [1977], and *Aleksandrova* [1988].)

classification per se. Returning to the work of *Gorodkov* [1939], he stated, "If some special zone must be assigned to the arctic, let it be the zone of the polar deserts." He pointed out that northward the podzolic process pinches out in the Polar Deserts and is replaced by a desertlike process. However, in the poorly drained locations, the gley process continues throughout the northern limits of ice-free land. *Rozov* [1956], *Ivanova* [1956], and *Ivanova and Rozov* [1962] used the term "Polar Desert" in their classification schemes (the actual translation can be termed Arctic or Arctic Desert rather than Polar Desert).

Fedoroff [1966], basing his observations on conditions in Svalbard, designated the well-drained sites in a broad sense as having polar soils (Arctic Polygonal soils).

Earlier a proposal by this writer for classifying soils of the High Arctic considered the gamut of soils along with a "wetness factor" plus the patterned ground form (Table 4).

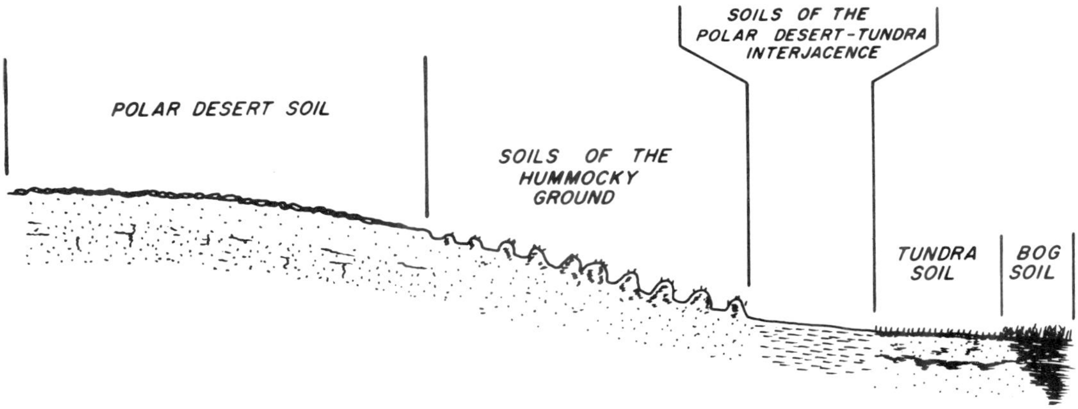

Fig. 3. Idealized cross section of soils within the Polar Deserts. Only the higher ground has Polar Desert (well-drained) soil; the other soils have some form of impeded drainage.

The U.S. Department of Agriculture (USDA) uses the terms "Pergelic Cryorthent" (frozen–main variety–recent) and "Pergelic Cryopsamment" (frozen-sandy-recent) for depicting sandy Polar Desert soils [*USDA*, 1975]. The Canadian Department of Agriculture (CDA) uses the terms "Regosolic Static Cryosol" and "Regosolic Trubic Cryosol" for the Polar Desert soil with the degree of frost churning being the factor used to separate the two [*Clayton et al.*, 1977]. The Soil Map of the World, published by the Food and Agriculture Organization–United Nations Educational, Scientific, and Cultural Organization [*FAO-UNESCO*, 1974], uses the term "Gelic Regosol" for Polar Desert soils. None of the above reports, however, mention conditions in Antarctica. The USDA, CDA, and FAO-UNESCO, using different nomenclature, provide for the desertlike conditions in their classifications but none recognize sets of pedogenic processes in a strict sense. Instead, they mainly emphasize temperature (including attendant frost processes) and the raw, earthy nature of the soil. This procedure is similar to the Arctic Hamada Råmark designation of *Kubiëna* [1953].

TABLE 4. A Proposed Classification for Soil of the Polar Desert

First Order	Second Order	Third Order	Fourth Order	Fifth Order
Polar Desert Soil Zone	Well-drained soils Polar Desert soil Arctic Brown soil* Mineral gley soils Upland Tundra Meadow Tundra Soils of the hummocky ground Soils of the Polar Desert–Tundra interjacence Organic soils Bog soils Other soils Regosols Lithosols Soils of the solifluction slopes (may be a form of gley soil but usually well drained) Rendzina†	Separations based on textural and mineral properties of the parent material, etc.	Soil type + wetness factor (applies mainly to Tundra and Bog soils)	Soil type + wetness factor + patterned ground

Source: *Tedrow* [1977] with modifications.
**Tedrow and Hill* [1955].
†*Tedrow and Walton* [1977].

Quaternary Events

In comparing Cold Desert soils of Antarctica to Polar Desert soils of the High Arctic, not only do climatic factors have to be considered, but the totality of facts relating to landscape evolution in the two sectors also needs to be properly equated. As a whole, the ice-free landscapes of the Antarctic are much older chronologically than are those in most of the High Arctic, with frost action, abrasion, aeolian activity, and other processes modifying the soil characteristics in both localities.

Isostatic Rebound in the High Arctic

A significant percentage of the ice-free land in the High Arctic was depressed below sea level during the Pleistocene epoch. The postglacial marine limit is currently about 400 ft (122 m) on Cornwallis Island [*Andrews*, 1973], 500 ft (152 m) in the Ellesmere–Axel Heiberg complex [*Andrews*, 1973], 300–350 ft (91–107 m) on Bathurst Island [*Blake*, 1964], 350 ft (107 m) in northern Greenland [*Davies*, 1961], 100 ft (30 m) in Svalbard [*Hoppe*, 1970], and 110 ft (34 m) in Franz Josef Land and surrounding areas [*Saks and Strelkov*, 1961]. The result of the postglacial uplift has had a major impact on soil development in that the more recently emerged areas lack maturely developed soils. In comparing soil processes in Antarctica with the well-drained ones within the High Arctic, the selection of sites is therefore critical because the chronologically older landscapes of the latter have greater Cold Desert soil affinities than do the more recently emerged lands. This condition is especially well exemplified on Bathurst Island.

Another factor affecting Polar Desert soil development on the more recently emerged lands is the nature of the substrate itself. In some situations the emerged lands may consist of little more than cliffs with virtually no soil cover. But where there is a constructional landform present, the nature of the substrate is of critical importance with the open-textured materials subject to greater soil development. Further, in some situations the emerged landscape may consist of little more than organic-derived material with the soil having boglike features [*Walton and Tedrow*, 1985; *Tedrow*, 1986]. The flat, organic-dominated soils result from a combination of present-day hydric vegetation mixed with the older plant debris from shallow marine waters.

Glaciation and Soil Development in the High Arctic

Prest [1976] showed that in the Queen Elizabeth Islands, the eastern sector of which includes Ellesmere Island and Devon Island, the ice-free areas have existed as such for the past 9000 years or so. In the western sector, which includes Prince Patrick and Melville islands, the land has been free of ice for the past 10,000 to 13,000 years. Glacial activity on the extreme western edge of Beaufort formation on Prince Patrick and Borden islands remains somewhat uncertain [*Craig and Fyles*, 1960]. The margins of northern Greenland have been free of ice for some 8500 years [*Weidick*, 1972]. In Svalbard, Franz Josef Land, and Severnaya Zemlya, the time frame approximates 8500 years [*Hoppe*, 1970; *Grosswald*, 1972]. In making a comparison of Antarctic soils to those of the High Arctic, it is therefore necessary to consider the overall time factor in soil development.

Soil Formation on Limestone

High-purity limestone within the High Arctic presents a special situation with respect to soil development [*Cruickshank*, 1971; *Tedrow*, 1978]. Limestone rocks tend to have lichen-free pitted and grooved surfaces but are otherwise unweathered. Silicate rocks, however, may be largely covered with lichens. Apparently, solution of limestone is at a sufficiently rapid rate to inhibit lichen colonization, but silicate rock surfaces are far more stable from the processes of direct solution. *Beschel* [1970] stated that rock deserts made up of limestone in the Queen Elizabeth Islands are more barren than the nonlimestone ones, because surfaces weather fast enough that epipetric lichens cannot become attached. Apparently the solution effect of the dissolving limestone carries over into soil formation. In situations in which the soil forms on high-purity limestone there is considerable silt and clay produced, which results in a matrix favorable for higher moisture contents, which, in turn, induces more intensive frost action. The soil itself consists of a dull gray silty clay with virtually no evidence of genetic horizonation. Polar Desert soil formed on silicate matrices will tend to develop better defined horizonation. High-purity limestone rock is present in much of the Canadian Arctic archipelago, Svalbard, Novaya Zemlya, Severnaya Zemlya, and elsewhere [*Tedrow*, 1978].

COMPARISON OF SOIL-FORMING FACTORS IN THE COLD DESERTS OF ANTARCTICA WITH THOSE OF THE HIGH ARCTIC POLAR DESERTS

Since the time of Dokuchaev, pedologists have generally agreed that the soil-forming factors are parent material, relief, climatic factors, biotic factors, and time. Accordingly, a functional organic component is generally considered necessary for the formation of a true soil. Because nearly all Antarctic soils are devoid of an effective organic component, one may question whether use of the term "soil" in Antarctica is justified. Not only are most Antarctic soils virtually free of humus but, in addition, a few apparently are abiotic [*Boyd et al.*, 1966; *Cameron*, 1969]. Though use of the term "soil" in Antarctica is somewhat moot, all investigators who have actually worked in Antarctica have used the term in a positive manner. Figure 4 shows a Cold Desert soil from Antarctica. The soil has a well-developed desert pavement and a dry aspect, but there is little visible horizonation.

Fig. 4. Cold Desert soil (Ahumic soil of the Cold Desert) from Wright Valley, south Victoria Land, Antarctica. The ruler shown here measures ~16 cm.

Fig. 5. Polar Desert soil formed on the Beaufort formation, Prince Patrick Island. The tape shown in the photograph measures ~60 cm.

Comparing the Cold Desert soil as shown in Figure 4 to the Polar Desert soil shown in Figure 5, gross appearances of the two are quite similar. Both soils are deep and well drained and have barely discernible genetic horizonation. A desert pavement is evident in both soils. The Polar Desert soil has a few irregular pockets of organic staining in the uppermost horizon, a condition not evident in the Cold Desert soil. Since there are a number of morphologic affinities between Cold Desert soils and Polar Desert soils, it is appropriate to compare the various processes leading to or associated with the two soils.

Rock Weathering

Rock disintegration is common throughout much of the Antarctic landscape. *Markov* [1956], *Avsiuk et al.* [1956], *Glazovskaia* [1958], and *Markov et al.* [1968] described shells of exfoliation and desquamational forces acting upon the rocks and the presence of weathered rock slabs. Pitted and fretted rock surfaces are quite common [*Nichols,* 1953]. *Campbell and Claridge* [1987] describe dolerites splitting along cleavage planes and argillite fracturing. Such deep-seated weathering processes do not appear to be very evident in the High Arctic. Rock weathering in the Cold Deserts is considered to result from a combination of processes such as freezing and thawing, movement of salts along rock temperature gradients, salt wedging, and others. There is a marked granular disintegration of boulders within Antarctic regoliths, especially those having coarse-textured rocks such as granite, dolerite, gneiss, and sandstone [*Campbell and Claridge*, 1987]. *Kelley and Zumberge* [1961] found that in the weathering of quartz diorite, the major chemical changes were related to iron conversion and limonite formation. *Tasch and Gafford* [1969] described slabs of argillite weathering in which the outer weathered zone was lighter in color with a marked decrease in organic carbon. Further, clay minerals of the inner zone consisted of unweathered illite, but the outer zone contained a weathered illite-montmorillonite mixed-layer mineral.

Wind Action

Wind action in Antarctica affects the landscape in the form of cavernous weathering of bedrock, ventifaction, and pitting of the surfaces of boulders and gravel [*Campbell and Claridge*, 1987]. Apparently, wind activity not only results in sand abrasion of surface rocks, but there is also good evidence of ice abrasion. Airborne ice crystals at extremely low temperatures, as exemplified by conditions in Antarctica, may have a hardness approximating that of feldspar or quartz.

Electron micrographs of samples from Beacon Valley, Antarctica, showed a high degree of chemical and mechanical weathering of sand grains, but samples from Wright Valley showed very little weathering [*Kumai et al.*, 1978].

In the High Arctic, evidence of abrasion from blowing sand (and especially ice crystals) does not appear to be as severe as in Antarctica, but there are locations where abraded bedrock is present. *Fristrup* [1952, 1952–1953] described wind erosion in the Polar Deserts of northern Greenland. There are also indications of wind-abraded rock in the western Canadian archipelago.

Desert Varnish and Desert Pavement

In Antarctica there is considerable evidence of desert varnish being present in the form of iron-enriched rock surfaces [*Glazovskaia,* 1958]. Such conditions appear to be far less common in the Polar Deserts. Desert pavement is widespread in Antarctica and can be considered a part of the soil profile. *Lindsay* [1973] showed that the grain size of the desert pavement increases with time. Desert pavement is also common in the High Arctic, especially on the Beaufort formation of northern Canada as well as on some stream outwash.

Salt Formation and Composition

Large quantities of salts are widely distributed throughout Antarctic soils, sediments, rocks, and lakes. The salts result from arid environment processes. There are also salt contributions from past marine transgression over some inland areas. Aerosols, salt spray, melt from glaciers, and possibly other sources contribute to overall saline conditions, but most salts apparently result from arid environment processes. *Campbell and Claridge* [1987] stated that salts occur as encrustations of efflorescences on rocks, soil surfaces, and under flat stones, as well as the soil itself. Although composition of the salts varies considerably from place to place, the salts generally consist of chlorides, sulfates, and nitrates of sodium, potassium, calcium, and magnesium. Gypsum crusts and gypsum-carbonate minerals are commonly distributed in Antarctic soils. Origin, composition, and distribution of the salts were discussed at length by *Glazovskaia* [1958], *Vornonov and Spiro* [1965], *Jones and Faure* [1968], *Black* [1969], *Tedrow* [1977], *Bockheim* [1980], *Ugolini* [1986*b*], *Campbell and Claridge* [1987], and many others.

In the High Arctic an accumulation of salts, mostly in the form of thenardite, has been reported [*Tedrow et al.,* 1968]. Soviet investigators have, for many years, reported the presence of "salt-fading crusts" (thernardite) throughout the polar sectors. After several rainless days with strong winds and low humidity, vegetative-free areas begin to acquire a grayish cast and, if dry conditions persist, these areas become covered with whitish encrustations of thenardite. Compared with Antarctica, however, salt accumulations in the High Arctic are much less. This condition was reported by *Tedrow* [1966, 1977], *Bockheim*

[1980], *Ugolini* [1986*b*], and others. *Bockheim* [1980] reported that electrical conductivity readings in both Cold Desert soils and Polar Desert soils increase with age, but overall readings from the former are much higher. In the High Arctic some of the salt deposits are from latent sources, as exemplified by alkali-saline lake clays, many of which have impeded drainage.

Moisture Regimes

Field measurements of moisture contents in both Cold Desert and Polar Desert soils are rather fragmental, but there have been enough data collected to make some preliminary statements. *Cameron and Conrow* [1969] reported a range of moisture from 0.51% to 3.01% in numerous locations in south Victoria Land. Relative humidity was found to approximate 40% at a depth of 5 cm, and it increased to about 90% at a depth of 30 cm. Cold Desert soils are generally in an extremely dry condition during summer (austral) months, an obvious condition to one digging a soil pit and encountering dusty conditions down to the frost table. *Ugolini and Anderson* [1973] stated that even though Antarctic ground temperatures are well below 0°C, the soils remain loose and unconsolidated and summer thaw seldom results in the appearance of liquid water. *Ugolini* [1965] reported the moisture content to be usually less than 1% in a Cold Desert soil having a sand content of about 99%. There was a small but measurable diurnal range of soil moisture.

Campbell and Claridge [1982] stated that moisture in Cold Desert soils is usually in the form of small ice crystals, vapor, and thin water films. The broad climatic differences in Antarctica are reflected in terms of moisture supply in the soils, which in turn enabled Campbell and Claridge to integrate moisture differences into their soil classification scheme.

Of the scant amount of snowfall occurring in Antarctica, probably most sublimates rather than melts. *Campbell and Claridge* [1987] cited individual situations, however, in which the meltwater from some snow patches moistens downslope positions. They reported greater moisture contents in the saline soils located in hollows and depressions. In such situations the soils are also more highly weathered. Some of the wet appearances of the saline deposits are associated with the deliquescence induced by the soluble salts.

The moisture regime of Polar Desert soils is in marked contrast to that of the Cold Deserts. Polar Deserts generally receive about 250 mm of snow during the winter months. There is considerable redistribution of the snow cover from wind activity with ridges and other exposed positions, those where Polar Desert soils are most likely to occur being largely snow-free. During early summer the thawing of seasonally accumulated ground ice results in a wet condition forming in the newly thawed soil. It is not uncommon for some Polar Desert soil sites to also receive runoff from summer snowmelt. On medium-textured soils the surface horizons have the appearance of raw, water-saturated mud; but on sands and gravels, especially those with desert pavement, soil conditions may be quite firm. As the summer progresses, the upper soil layers become quite dry, but the recently thawed plane immediately above the receding frost table remains moist to wet. By midsummer the uppermost soil layers generally become quite dry and, especially with medium- to heavy-textured soils, there is a tendency for the soil to effloresce. But with the appearance of summer rain, drizzle, fog, or even periods of high humidity the thin veneer of salt crusts may disappear. *Gorodkov* [1939] reported that some of the moistening effect of Polar Desert soils is also from atmospheric condensation. Comparing the moisture regime of Cold Desert soils to that of Polar Desert soils, the latter has a rather complex set of conditions, one in which the soil is subject to a wetting effect at any time throughout the summer months, a condition unheard of in the Cold Deserts except possibly at the northern end of the Antarctic Peninsula.

Temperature Regimes

During the austral summer the gravelly coarse-textured Antarctic soils will be in an unfrozen state ranging from a depth of a few centimeters to as much as 50 cm or even more. At lower depths the soils may have a negative temperature, but because of the low moisture content they will remain in an unconsolidated or poorly consolidated state, loose and granular.

Ugolini [1965] reported daily summer temperature amplitudes of as much as 10°C in the uppermost (2 cm) soil layer with the amplitude decreasing with depth. *Cameron* [1969] recorded daily summer soil temperature ranges of 15° to 18°C, with a marked decrease with depth. Various investigators have projected that near-surface soil temperatures are related more to solar radiation than to ambient temperatures. It is also important to mention the temperatures of rock surfaces in Antarctica, which on bright, sunny days, with little wind activity, may be in the 25° to 42°C range. It was the high surface temperatures of rocks in the Mirnyy sector that led Markov to project salt migration along the temperature gradients of the rocks.

Midsummer soil temperature measurements compiled from the Cold Deserts and the Polar Deserts are generalized in Figure 6. The figure does not address the question of the effects of the substrate, diurnal variation, and other individual site characteristics.

Biotic Elements

The organic component in Cold Desert soils of Antarctica is not highly significant insofar as the soil system is concerned. *Cameron* [1969] reported a group of Cold Desert soils having a range of 0.002% to 0.24% organic carbon. Whether the organic matter

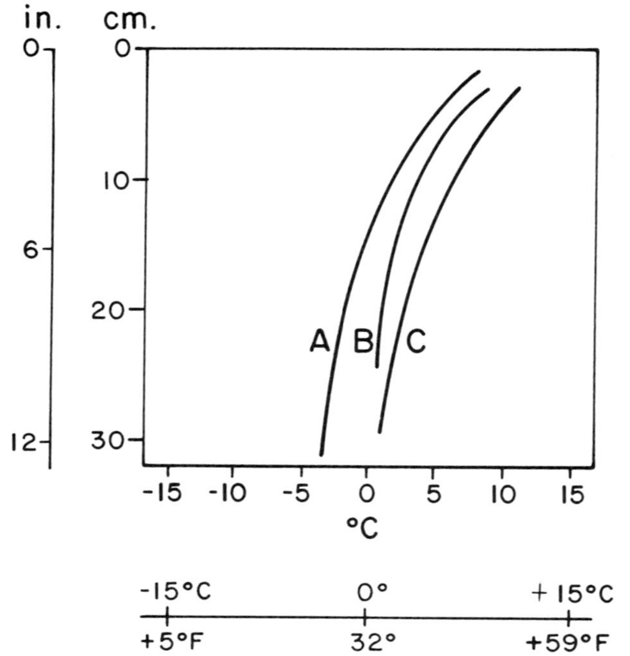

Fig. 6. A comparison of midsummer soil temperatures. Curve A: December soil temperature from south Victoria Land, Antarctica. The soil is described as Ahumic soil of the Cold Desert [*Tedrow and Ugolini*, 1966]. Curve B: July soil temperature from Franz Josef Land. The soil is described as Polar Desert [*Aleksandrova*, 1988]. Curve C: July soil temperature of Polar Desert soil from Inglefield Land, Greenland [*Tedrow*, 1970].

is totally from plant life or in part from bird droppings or material brought in by aeolian activity remains unknown. Also, some organic carbon may have been derived from weathered rocks [*Tasch and Gafford*, 1969]. *Boyd et al.* [1966], *Rudolph* [1966], *Janetschek* [1963], *Cameron* [1969, 1974], and others have outlined the abundance and ecology of terrestrial life-forms on the continent (Figure 7).

The organic component in Polar Desert soils plays far more integral part of the soil system than is the case in Cold Desert soils. The organic matter content of the uppermost few inches may be as much as 5% or more, but on sandy, open-textured material it is generally at the 1% to 2% level. *Gorodkov* [1939] believed that a significant amount of the organic matter in Polar Deserts was derived from algae, including diatoms. *Whittaker* [1970] reported that primary productivity approximates 3 g m^{-2} yr^{-1} in the Polar Deserts compared with 800 g in the boreal forest and 1300 g in the temperate forests. Apparently, the organic matter in Polar Desert soils is generally quite old. A ^{14}C date from the surface soil in northern Greenland yielded an age of 3300 ± 110 years B.P. [*Tedrow*, 1970].

Figure 7 shows how life spectra pinch out as the environment becomes harsher. Thus within the Polar Deserts the groups of organisms would be typified by the right-hand side of the diagram, but the Cold Desert conditions would be exemplified more to the left.

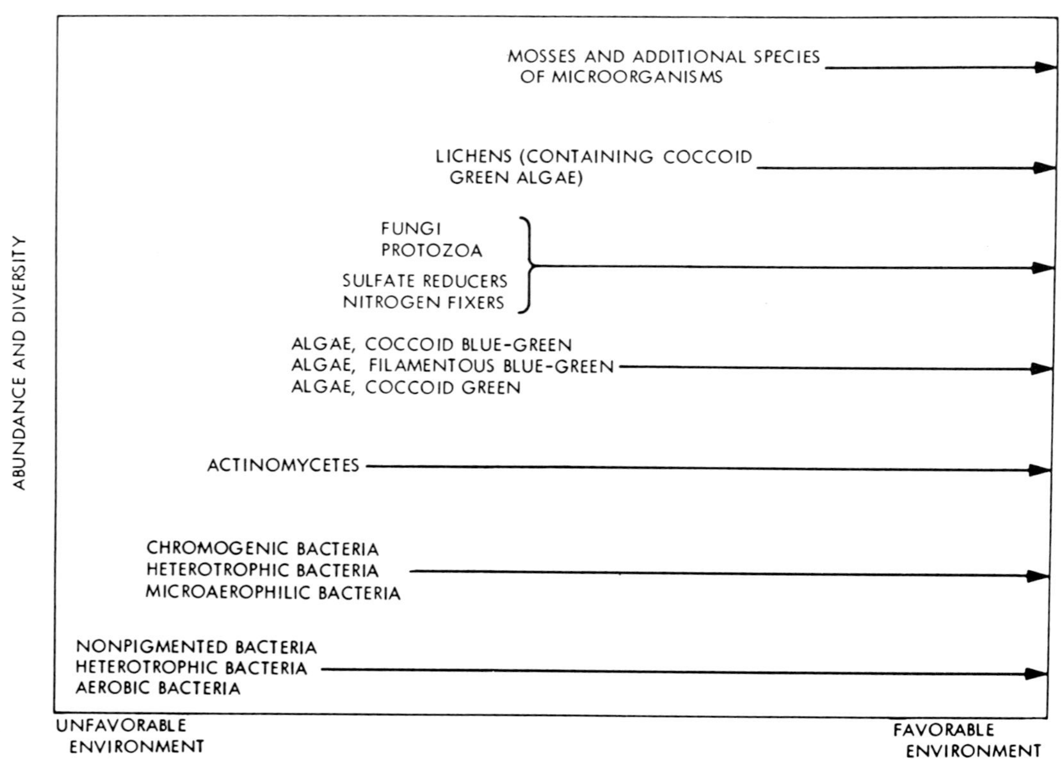

Fig. 7. Variability of plant population density and diversity of organisms with ecologic factors in the dry valleys, Victoria Land, Antarctica. Conditions on the extreme right approach conditions of the High Arctic. (Diagram courtesy of R. L. Cameron.)

The Time Factor

The importance of the time factor as it relates to soil development in both the Cold Deserts and the Polar Deserts has been recognized by virtually all pedologists. *McCraw* [1960] recognized greater soil development with increased age, and since his work was published there have been many investigations relating to the "browning" effect, formation of carbonate and gypsum layers, and weathering of rocks, among others, with time. In the Polar Deserts the soils tend to acquire a somewhat yellowish color with increased age, and carbonate minerals form but the latter process remains weakly expressed [*Everett*, 1969].

Recently, a number of important investigations related soil development to absolute chronology. *Bockheim* [1980] reported that pedogenic horizons in both Cold Desert soils and Polar Desert soils are limited to desert pavement and to a weak B horizon in both soils. In addition, solum thickness increases with age, but because of more moisture in Polar Desert soils the solum thickness is greater than in Cold Desert soils. *Bockheim* [1980] also provided some quantitative guidelines as to rates of soil formation in the two soil zones. *Bockheim* [1982] extended his studies into the Transantarctic Mountains, where he reported increased thickness and depth of the solum and related features with time. The problem of soil development in Antarctica on a time basis was reviewed in detail by *Campbell and Claridge* [1987].

COMPARISON OF PROPERTIES OF COLD DESERT SOILS WITH THOSE OF POLAR DESERT SOILS

Comparing "modal" Ahumic soil of the Cold Desert (Figure 8) with Polar Desert soil (Figure 9) will show a number of morphologic features in common. Both have a desert aspect, including desert pavement, a mineral appearance, and a small organic component. They are also subjected to low temperatures, and they have cryogenic features and possibly other related characteristics. In most cases, however, there are a number of quantitative differences between the two. In addressing the above problem, site selection is critically important, perhaps more important in soil morphology than in any other branch of science. If one searches for a perceived set of soil properties in the Cold Deserts of Antarctica and then searches for a similar set in the Polar Deserts of the High Arctic, such conditions could probably be located. An investigator would, however, be knowingly practicing eclecticism. If one had the same type of substrate within the same topographic setting, the same age, etc., under the two mentioned regions, there would probably be recognizable differences relative to individual soil properties. In addition, if a Polar Desert soil were examined on the north coast of Greenland in late June and again in mid-August, descriptions and subsequent laboratory findings would probably be somewhat different. On the former date the frost table (tjaele) might be at a depth of 450 mm, and the soil might be wet with no efflorescence. But by late summer the frost table would probably be much deeper, and possibly the soil surface would be veneered with salt.

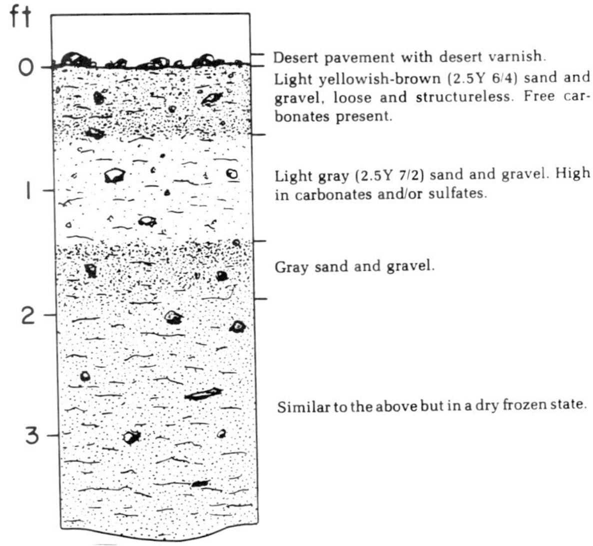

Fig. 8. Idealized Cold Desert soil (Ahumic soil of the Cold Desert in Antarctica). (Scale: 1 foot = 30.48 cm)

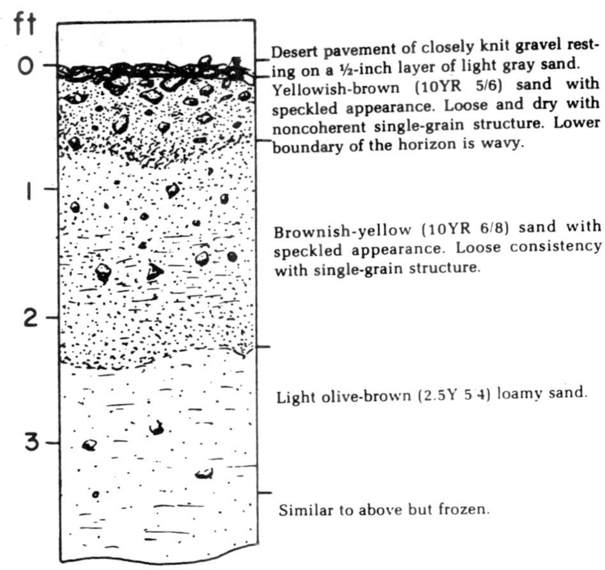

Fig. 9. Idealized Polar Desert soil of the High Arctic (Prince Patrick Island). (Scale: 1 foot = 30.48 cm)

A number of investigators have addressed the complexities of comparing Cold Desert soils with Polar Desert soils [e.g., *Tedrow*, 1968, 1977; *Bockheim*, 1980; *Ugolini* 1986*a*, *b*; *Campbell and Claridge*, 1987]. Most discussions centered upon a mix of data collection, field observations, and a "statement of the problem" with few definite conclusions.

A nearly complete spectrum of clay minerals was reported in the Cold Desert soils of Antarctica. Various light mica forms such as illite and hydrous mica appear to dominate in most situations, but smectite-montmorillonite, vermiculite, chlorite, and, in some situations, kaolinite are common. A few less common clay mineral forms were also reported [e.g., *Claridge*, 1965; *Everett*, 1971; *Bockheim*, 1980; *Ugolini and Jackson*, 1982; *Ugolini*, 1986*a, b; Campbell and Claridge*, 1987].

It appears that most of the clay minerals are of an allogenic nature, but there has been some low-order diagenesis, as evidenced by the presence of interlayered hydrous micas and Al-chlorite minerals. *Bockheim* [1980] suggested a muscovite mica > vermiculite weathering sequence, with further weathering to montmorillonite and chloritized montmorillonite. Various other intermediate weathering forms have also been reported. Quartz and feldspar persist in the clay-size fractions.

In Polar Desert soils, the clay minerals consist mainly of vermiculite, illite, and kaolinite, as well as smaller quantities of montmorillonite and chlorite plus various interstratified minerals [e.g., *Everett*, 1968; *Tedrow*, 1977; *Bockheim*, 1980; *Ugolini*, 1986*a*, *b*]. *Bockheim* [1980] proposed a muscovite mica > hydrated mica > vermiculite > kaolinite weathering sequence in Polar Desert soils. Clay-size quartz and feldspar are also present and, according to Bockheim, feldspar weathers directly to kaolinite.

Total elemental composition of soil, once a major method for depicting soil processes, is now seldom used in pedologic research except in special situations. *Tedrow and Thompson* [1969] analyzed several soils from the Cold Deserts of Antarctica and the Polar Deserts of the High Arctic, however, for general baseline information. The data showed a degree of surface leaching in the Polar Desert soil but not in the Cold Desert soil. Also, of the 15 trace elements determined, few trends relating to pedogenesis were found. There were, however, unusually high values of strontium in the Antarctic samples.

Cold Desert soils usually show an alkaline reaction, with *p*H values generally ranging from 7 to 9, but values as low as 4.6 and as high as 10 have been reported [*Claridge*, 1965]. Comparing *p*H values of Cold Desert soils to those of Polar Deserts becomes somewhat complex, but as a whole the Cold Desert soils have higher *p*H values [*Campbell and Claridge*, 1987]. The variation of *p*H values among Polar Desert soils is related primarily to the substrate and the time element. For example, Polar Desert soils on the Beaufort formation, a very acid material on Prince Patrick Island, were found to be as low as *p*H 4; the soils on nearby carbonate-bearing sandstone were found to be alkaline. *Bockheim* [1980] stated that Polar Desert soils at 74°N and most Cold Desert soils are mildly alkaline, but at 66°N (the Subpolar Desert zone of *Tedrow* [1977]) the soils are strongly to medium acid.

The degree of salinity in the Cold Desert soils is generally far greater than in Polar Desert soils, but some saline features exist also in the Polar Deserts. It appears that the salinity factor in the Polar Desert soils is more of a transient property with a marked seasonal amplitude—a condition probably quite different than in Cold Desert soils. *Karavaeva and Targulyan* [1977] recognized an independent group of saline soils within the Polar Deserts of Eurasia.

It has been shown that calcification (pedogenic carbonate formation) and salinization operate in both Cold Desert soils [e.g., *Campbell and Claridge*, 1987] and in Polar Desert soils [*Tedrow*, 1968, 1970; *Everett*, 1968; *Sletten*, 1988], but the intensities of the processes are greater in the former.

Easily reducible iron values in Cold Desert soils increase with age of the landscape. *Campbell and Claridge* [1987] reported that the soils of weathering stage 1 (initial) have 0.1–0.3% Fe_d. Apart from the work of *Campbell and Claridge* [1987] showing an increase in quantity of easily reducible iron with age in Antarctic soils, there appear to be no specific distribution patterns (discounting soluble salt forms) within the various soils or any regular differences between Cold Desert soils of Antarctica and Polar Desert soils of the High Arctic [*Tedrow et al.*, 1968; *Bockheim*, 1980].

CONCLUSIONS

Cold Desert soils of Antarctica have a number of properties in common with the Polar Desert soils of the High Arctic. Both have desertlike features such as a mineral appearance and a desert pavement and are subjected to low temperatures and possibly other similarities. Whether soils of the two sectors should be combined into one classification unit is primarily a judgment decision, one involving semantics as well as a comparison of selected soil properties. It appears, however, that with the present state of knowledge there are enough distinguishing characteristics to tentatively continue to separate the two. This position is based primarily on the differences in moisture and temperature regimes, the organic component, and the leaching potential within the two sectors. Moreover, Polar Desert soils have a number of critical soil properties, such as moisture content and the degree of leaching and salt accumulation, that have a high order of seasonal variability—a condition of minimal importance in the Cold Desert soils.

Acknowledgments. This project was supported in part by National Science Foundation grant G-17212 and various Office of Naval Research grants.

REFERENCES

Aleksandrova, V. D., *Vegetation of the Soviet Polar Deserts,* translated from Russian by D. Love, p. 3, Cambridge University Press, New York, 1988.

Andrews, J. T., Maps of the maximum postglacial marine limit and rebound for the former Laurentide ice sheet, *Arct. Alp. Res., 5,* 41–48, 1973.

Avsiuk, G. A., K. K. Markov, and P. A. Shumskiy, Cold Deserts in Antarctica (in Russian), *Izv. Akad. Nauk SSSR, Ser. Geogr.,* no. 4, 16–25, 1956.

Behling, R. E., Pedogical development on moraines of the Merserve Glacier, Antarctica, Ph.D. thesis, 216 pp., Ohio State Univ., Columbus, 1971.

Beschel, R. E., The diversity of tundra vegetation, in *Productivity and Conservation in Northern Circumpolar Lands, New Ser.,* vol. 16, edited by W. A. Fuller and P. G. Kevan, pp. 85-92, International Union for Conservation of Nature and Natural Resources, Morges, Switzerland, 1970.

Black, R. F., Saline discharges from Taylor Glacier, Victoria Land, Antarctica, *Antarct. J. U.S., 4,* 89–90, 1969.

Blake, W., Jr., Preliminary account of the glacial history of Bathurst Island, Arctic archipelago, *Pap. Geol. Surv. Can.,* no. 64-30, 8 pp., 1964.

Blanck, E., Ein Beitrag zur Kenntnis arktischer Boden, insbesondere Spitzbergen, *Chem. Erde, 1,* 421–476, 1919.

Bliss, L. C., *Truelove Lowland, Devon Island, Canada: A High Arctic Ecosystem,* 714 pp. University of Alberta Press, Edmonton, Alta., 1977.

Bockheim, J. G., Relative age and origin of soils in eastern Wright Valley, Antarctica, *Soil Sci., 128,* 142–152, 1979.

Bockheim, J. G., Properties and classification of some desert soils in the coarse-textured drift in the Arctic and Antarctic, *Geoderma, 24,* 45–69, 1980.

Bockheim, J. G., Properties of a chronosequence of Ultraxerous soils in the Trans-antarctic Mountains, *Geoderma, 28,* 239–255, 1982.

Bockheim, J. G., and F. C. Ugolini, A review of pedogenic zonation in well drained soils of the southern circumpolar region, *Quat. Res. N.Y., 34,* 47–66, 1990.

Bockheim, J. G., S. C. Wilson, and J. W. Leide, Soil development in the Beardmore Glacier region, Antarctica, *Soil Sci., 149,* 144–157, 1990.

Bovis, M. J., and R. G. Barry, A climatological analysis of north polar desert areas, in *Polar Deserts and Modern Man,* edited by T. L. Smiley and J. H. Zumberge, pp. 23–31, University of Arizona Press, Tucson, 1974.

Boyd, W. L., J. T. Staley, and J. W. Boyd, Ecology of soil microorganisms of Antarctica, in *Antarctic Soils and Soil Forming Processes, Antarct. Res. Ser.,* vol. 8, edited by J. C. F. Tedrow, pp. 125–159, AGU, Washington, D. C., 1966.

Cameron, R. L., Cold desert characteristics and problems relevant to other arid lands, in *Arid Lands in Perspective,* edited by W. G. McGinnis and B. J. Goldman, pp. 167–205, University of Arizona Press, Tucson, 1969.

Cameron, R. L., Application of low-latitude microbial ecology to high-latitude ecology, in *Polar Deserts and Modern Man,* edited by T. L. Smiley and J. H. Zumberge, pp. 71–90, University of Arizona Press, Tucson, 1974.

Cameron, R. L., and H. P. Conrow, Soil moisture, relative humidity and microbial abundance in dry valleys of Southern Victoria Land, *Antarct. J. U.S., 4,* 23–28, 1969.

Campbell, I. B., and G. G. C. Claridge, A classification of frigic soils—The zonal soils of the Antarctic continent, *Soil Sci., 107,* 75–85, 1969.

Campbell, I. B., and G. G. C. Claridge, The influence of moisture on the development of the soils of the Cold Deserts of Antarctica, *Geoderma, 28,* 221–238, 1982.

Campbell, I. B., and G. G. C. Claridge, *Antarctica: Soils, Weathering Processes and Environment,* 368 pp. Elsevier, New York, 1987.

Claridge, G. G. C., The clay mineralogy and chemistry of some soils from the Ross Dependency, Antarctica, *N.Z. J. Geol. Geophys., 8,* 186–220, 1965.

Clayton, J. S., et al., *Soils of Canada,* vol. 1, pp. 23-26, Canadian Department of Agriculture, Ottawa, 1977.

Craig, B. S., and J. G. Fyles, Pleistocene geology of Arctic Canada, *Geol. Surv. Can.,* paper 60-10, 22 pp., 1960.

Cruickshank, J. G., Soils and terrain units around Resolute, Cornwallis Island, *Arctic, 24,* 195–209, 1971.

Davies, W. E., Glacial geology of northern Greenland, *Polarforschung, 5,* 95–102, 1961.

Egede, H. P., *A Description of Greenland,* translated from Danish, 220 pp., C. Hitch, London, 1745.

Everett, K. R., Soil development in the Mould Bay and Isachsen areas, Queen Elizabeth Islands, N.W.T., Canada, *Rep. 24,* 75 pp., Inst. of Polar Stud., Ohio State Univ., Columbus, 1968.

Everett, K. R., Pedology of the Trinity Peninsula and offshore islands, *Antarct. J. U.S., 4,* 138, 1969.

Everett, K. R., Soils of the Meserve glacier area, Wright Valley South Victoria Land, Antarctica, *Soil Sci., 112,* 425–438, 1971.

Everett, K. R., A survey of soils in the region of the South Shetland Islands and adjacent parts of the Antarctic Peninsula, *Rep. 58,* 44 pp., Inst. of Polar Stud., Ohio State Univ., Columbus, 1976.

FAO-Unesco, Soil map of the world, Food and Agric. Organ.–U.N. Educ., Sci., and Cult. Organ., Rome, 1974.

Fedoroff, N., Les sols du Spitsberg Occidental, in *Spitsberg 1964,* chap. 10, pp. 111–128, Centre National de la Recherche Scientifique Programme Coopératif, Lyons, France, 1966.

Fristrup, B., High Arctic desert, *Int. Geol. Congr. Proc., 7,* 91–99, 1952.

Fristrup, B., Wind erosion within the Arctic deserts, *Saertr. Geogr. Tidsskr., 52,* 51–65, 1952–1953.

Glazovskaia, M. A., Weathering and primary soil formation in Antarctica (in Russian), *Nauchr. Dokl. Vyssh. Shk. Geol. Geogr. Nauki, 1,* 63–76, 1958.

Gorodkov, B. N., *Vegetation in the Tundra Zone of the U.S.S.R.* (in Russian), 142 pp., Academy of Sciences of the USSR, Moscow, 1935.

Gorodkov, B. N., Peculiarities of the Arctic topsoil (in Russian), *Izv. Gos. Geogr. O-Va., 71,* 1516–1532, 1939.

Grosswald, M. G., Glacier variations and crustal movements in northern European Russia in Late Pleistocene and Holocene times, in *Climatic Changes in Arctic Areas During the Last Ten Thousand Years, Acta Univ. Ouluensis Ser. A. (3) Geol.,* vol. 1, edited by Y. Vasary et al., pp. 205–223, Oulu University Library, Oulu, Finland, 1972.

Hoppe, G., The Wurm ice sheets of northern and Arctic Europe (in Polish), *Acta Geogr. Lodz., 24,* 205–215, 1970.

Ivanova, E. N., Classification of soils of the northern parts of the U.S.S.R. (in Russian), *Pochvovedenie,* no. 1, 70–78, 1956.

Ivanova, E. N., and N. N. Rozov, *Soil Geographical Zoning of the U.S.S.R.*, translated from russian by D. Davey, 479 pp., Academy of Sciences of the USSR, Moscow, 1962.

Janetschek, H., Biologie von Antarctika, *Ber. Naturwiss. Med. Ver. Innsbruck, Gams-Festschr., 53,* 235–246, 1963.

Jensen, H. I., Report on Antarctic soils, *Br. Antarct. Exped. 1907–1909 Sci. Invest. Rep. Geol. 6,* pp. 89–92, Heinemann, London, 1916.

Jones, L. M., and G. Faure, Origin of the salts in Taylor Valley, *Antarct. J. U.S., 3,* 177–178, 1968.

Karavaeva, N. A., and V. O. Targulyan, Autonomous soil formation in the north of Eurasia and America (in Russian), in *Problems of Soil Science,* pp. 301–309, Academy of Sciences of the USSR, Moscow, 1977.

Kelley, W. C., and J. H. Zumberge, Weathering of a quartz diorite at Marble Point, McMurdo Sound, Antarctica, *J. Geol., 69,* 433–446, 1961.

Korotkevich, Y. S., Polar deserts, *Sov. Antarct. Exped. Inf. Bull. Engl. Transl. 6*(65), 459–472, 1967.

Kubiëna, W. L., *The Soils of Europe,* 317 pp., T. Murby, London, 1953.

Kumai, M., D. M. Anderson, and F. C. Ugolini, Antarctic soil studies using a scanning electron microscope, *Proc. Int. Conf. Permafrost 3rd,* 106–112, 1978.

Lindsay, J. F., Ventifact evolution in Wright Valley, Antarctica, *Geol. Soc. Am. Bull. 84,* 1791–1798, 1973.

Liverovskij, J. A., Soils of the tundras of the northern regions (in Russian), *Tr. Polyarn. Kom. Akad. Nauk SSSR, 19,* 112 pp., 1934.

MacNamara, E. E., Active layer development and soil moisture dynamics in Enderbyland, East Antarctica, *Soil Sci., 108,* 345–459, 1969*a*.

MacNamara, E. E., Soils and geomorphic surfaces in Antarctica (in English), *Biul. Peryglacjalny, 20,* 99–320, 1969*b*.

Makeev, O. V., The soils of the polar climate regions, their major limitations for food production and the overriding climatic restraints, *Proc. Int. Soc. Soil Sci. 11th, 2,* 27–70, 1978.

Markov, K. K., Some facts concerning periglacial phenomena in Antarctica, preliminary report (in Russian), *Vestn. Mosk. Univ. Ser. Biol. Pochvoved. Geol. Geogr., 11 (1)* 139–148, 1956.

Markov, K. K., V. I. Bardin, V. L. Lebedev, A. I. Orlov, and I. A. Suetova, *The Geography of Antarctica,* translated from Russian by N. Kaner, edited by J. Sherman, 369 pp., Israel Program for Scientific Translations, Jerusalem, 1970. (Available from the U.S. Department of Commerce, Clearinghouse for Federal Scientific and Technical Information, Springfield, VA.)

McCraw, J. D., Soils of the Ross Dependency Antarctica, a preliminary note, *N.Z. Soc. Soil Sci. Proc., 4,* 30–34, 1960.

McCraw, J. D., Soils of the Taylor Dry Valley, Antarctica with notes on soils from other localities in Victoria Land, *N.Z. J. Geol. Geophys., 10,* 498–539, 1967.

Meinardus, W., Uber einige characteristiche Bodenformen auf Spitzbergen, *Sitzungsber. Naturhist. Ver. Preuss. Rheinlände Westfalens,* pp. 1–42, 1912.

Nichols, R. L., Geomorphology of Marguerite Bay, Palmer Peninsula, Antarctica, *Tech. Rep. 12,* 151 pp., *Off. of Nav. Res.,* Washington, D. C., 1953.

Nikiforoff, C. C., General trends of the desert type of soil formation, *Soil Sci., 43,* 105–132, 1937.

Nordenskjöld, O. G., and L. Mecking, *The Geography of the Polar Regions, Spec. Publ.,* no. 8, p. 74, American Geographical Society, New York, 1928.

Passarge, S., *Die Grundlagen der Landschaftskunde,* 588 pp., Hamburg, 1920.

Pastor, J., and J. G. Bockheim, Soil development on moraines of Taylor Glacier, Lower Taylor Valley, Antarctica, *Soil Sci. Soc. Am. J., 44,* 341–348, 1980.

Prest, V. K., Quaternary geology of Canada, in *Geology and Economic Minerals of Canada,* part B, edited by R. J. W. Douglas, pp. 675–764, Department of Energy, Mines, and Resources, Ottawa, Ont., 1976.

Rozov, N. N., Principles of elaborating a genetic classification of soils (in Russian), *Pochvovedenie,* no. 6, 76–83, 1956.

Rudolph, E. D., Terrestrial vegetation in Antarctica: Past and present studies, in *Antarctic Soils and Soil Forming Processes, Antarct. Res. Ser.,* vol. 8, edited by J. C. F. Tedrow, pp. 109–124, AGU, Washington, D. C., 1966.

Saks, V. N., and S. A. Strelkov, Mesozoic and Cenozoic of the Soviet Arctic, in *Geology of the Arctic,* vol. 1, edited by G. O. Raasch, pp. 48–67, University of Toronto Press, Toronto, Ont., 1961.

Sletten, R. S., The formation of pedogenic carbonates on Svalbard: The influence of cold temperatures and freezing, *Proc. Int. Conf. Permafrost 3rd,* 467–472, 1988.

Syroechkovsky, E. E., The role of animals in primary soil formation under conditions of the pre-polar region of the globe, as exemplified by the Antarctic (in Russian), *Zool. Zh., 38,* 1770–1775, 1959.

Tasch, P., and E. I. Gafford, Weathering of an Antarctic argillite: Field, geochemical and mineralogical observations, *J. Sediment. Petrol., 39,* 369–373, 1969.

Tedrow, J. C. F., Polar Desert soils, *Soil Sci. Soc. Am. Proc., 30,* 381–387, 1966.

Tedrow, J. C. F., Pedogenic gradients of the polar regions, *J. Soil Sci., 19,* 197–204, 1968.

Tedrow, J. C. F., Soil investigations in Inglefield Land, Greenland, *Medd. Groenl., 188*(3), 92 pp., 1970.

Tedrow, J. C. F., Polar soil classification and the periglacial problem, *Biul. Peryglacjalny, 22,* 285–294, 1973.

Tedrow, J. C. F., Soils of the High Arctic landscapes, in *Polar Deserts and Modern Man,* edited by T. H. Smiley and J. H. Zumberge, pp. 63–69, University of Arizona Press, Tucson, 1974.

Tedrow, J. C. F., *Soils of the Polar Landscapes,* 638 pp., Rutgers University Press, New Brunswick, N. J., 1977.

Tedrow, J. C. F., Development of Polar Desert soils, in *Quaternary Soils,* edited by W. H. Mahaney, pp. 413–425, Geological Abstracts, Norwich, England, 1978.

Tedrow, J. C. F., Factors affecting Polar Desert soil development in the High Arctic, *Biul. Periglacjalny, 31,* 275–282, 1986.

Tedrow, J. C. F., and D. E. Hill, Arctic Brown soil, *Soil Sci., 80,* 265–275, 1955.

Tedrow, J. C. F., and C. C. Thompson, Chemical composition of polar soils, *Biul. Periglacjalny, 18,* 169–181, 1969.

Tedrow, J. C. F., and F. C. Ugolini, Antarctic soils, in *Antarctic Soils and Soil Forming Processes, Antarct. Res. Ser.,* vol. 8, edited by J. C. F. Tedrow, pp. 161–177, AGU, Washington, D. C., 1966.

Tedrow, J. C. F., and G. F. Walton, Rendzina formation on Bathurst Island, *J. Soil Sci., 28,* 519–525, 1977.

Tedrow, J. C. F., P. F. Bruggemann, and G. F. Walton, Soils of Prince Patrick Island, *Tech. Pap. Arct. Inst. North Am. 44,* 81 pp., Univ. of Calgary, Calgary, Alta., Canada, 1968.

Ugolini, F. C., Soil investigations in the Lower Wright Valley, *Proc. Int. Conf. Permafrost 1st,* 55–61, 1965. (Available as NAS-NRC Publ. 1287 from the National Academy of Sciences, Washington, D. C.)

Ugolini, F. C., Antarctic soils and their ecology, in *Antarctic Ecology,* vol. 2, edited by M. W. Holdgate, pp. 673–692, Academic, New York, 1970.

Ugolini, F. C., Pedologic zonation in the well-drained soils of the Arctic regions, *Quat. Res. N.Y., 26,* 100–120, 1986*a*.

Ugolini, F. C., Processes and rates of weathering in Cold and Polar Desert environments, in *Rates of Chemical Weathering of Rocks and Minerals,* chap. 10, pp. 193–235, Academic, New York, 1986*b*.

Ugolini, F. C., and D. M. Anderson, Ionic migration and weathering in frozen Antarctic soils, *NASA Contract. Rep. CR-2283,* 24 pp., 1973.

Ugolini, F. C., and M. L. Jackson, Weathering and mineral synthesis in Antarctic soils, in *Antarctic Geoscience,* vol. 142, pp. 1101–1108, University of Wisconsin Press, Madison, 1982.

Ugolini, F. C., and R. L. Starkey, Soils and microorganisms from Mt. Erebus, Antarctica, *Nature, 211,* 440–441, 1966.

USDA, *Soil Taxonomy, Agric. Handb.,* no. 436, pp. 62, 86–87, U.S. Department of Agriculture, Washington, D. C., 1975.

Voronov, P. S., and N. S. Spiro, Composition of soluble salts in Quaternary terrigenic sediments of the central and western sectors of East Antarctica (in Russian), *Probl. Arkt. Antarkt., 15,* 5–10, 1965.

Walton, G. F., The High Arctic environment and Polar Desert soils, Ph.D. thesis, 479 pp., Rutgers Univ., New Brunswick, N.J., 1972.

Walton, G. F., and J. C. F. Tedrow, A soil pattern of Central Bathurst Island, Queen Elizabeth Islands, Canada, *Biul. Peryglacjalny, 30,* 127–139, 1985.

Weidick, A., Holocene shore lines and glacial stages in Greenland: An attempt at correlation. *Rapp. Groenl. Geol. Unders., 41,* 39 pp., 1972.

Whittaker, R. H., *Communities and Ecosystems,* 162 pp., Macmillan, New York, 1970.

(Received September 5, 1990;
accepted October 15, 1990.)

CONTRIBUTIONS TO ANTARCTIC RESEARCH II
ANTARCTIC RESEARCH SERIES, VOLUME 53, PAGES 19–31

ABLATION RATES OF THE ICE FIELDS IN THE VICINITY OF THE ALLAN HILLS, VICTORIA LAND, ANTARCTICA

GUNTER FAURE AND DAVID BUCHANAN

Department of Geological Sciences and Byrd Polar Research Center
Ohio State University, Columbus, Ohio 43210

The ablation rates on the ice fields west of the Allan Hills were determined by measuring the lowering of the ice surface, recorded by 265 wooden dowels implanted in the ice, at the end of the summer and again 2 years later. The results were used to estimate both annual and summer ablation rates at the stations distributed among the Elephant Moraine (76°17′S, 157°20′E), the Reckling Moraine (76°15′S, 158°40′E), and the Main ice field of the Allan Hills (76°47′S, 159°25′E). The average annual ablation rates increase toward the edge of the ice sheet from 4.1 ± 0.1 cm yr^{-1} at Elephant Moraine to 4.7 ± 0.2 cm yr^{-1} at Reckling Moraine and to 5.3 ± 0.1 cm yr^{-1} at the Allan Hills. The average summer ablation rates on the ice sheets at these localities are about five times faster than the average annual rates and increase toward the margin of the ice sheet even more than the annual rates. Rapid ablation during the summer can remove a layer of ice equal to the annual rate in about 70 days, provided that the ablation rate during the winter is assumed to be negligible. The average annual and summer ablation rates within the moraines are significantly lower than those of the open ice fields. The difference can be attributed to the presence of snow and rock clasts on the surface of the ice. The summer ablation rates in the moraines are from seven to eight times faster than the annual rates within the moraines. The greater increase of summer ablation rates in the moraines compared to the ice fields may be caused by solar heating of the rock clasts in the moraines. The present topography of the ice sheet at the Reckling Moraine is independent of the ablation rates. But within about 1 km of the ice margin at the Allan Hills, ablation rates vary inversely as the surface elevations and appear to exert a strong influence on the topography of the marginal zones of the ice sheet.

INTRODUCTION

The ice fields on the East Antarctic ice sheet northwest of the Allan Hills, Victoria Land, have yielded a large number of meteorite specimens [*Cassidy et al.*, 1977]. The accumulation of meteorites in these areas was explained by *Whillans and Cassidy* [1983] by means of an ice-flow model, according to which meteorites were transported englacially from the accumulation to the ablation zone of the ice sheet and were concentrated on its surface as a result of ablation of the ice, augmented by compressive ice flow and by meteorites falling directly into the ablation zone.

The ice fields on which meteorites have been found are associated with ice escarpments where the flow of the ice sheet is disturbed by the topography of the underlying bedrock [*Faure*, 1990]. The effect of bedrock topography on ice-flow patterns is most evident at two sites where large moraines composed of terrestrial rocks have formed. The moraines, known as the Elephant Moraine and the Reckling Moraine (Figure 1), consist of clasts composed of a variety of sandstones, dolerite, basalt, diamicton, and coal similar to the rocks that form the Transantarctic Mountains [*Taylor*, 1986; *Montello*, 1986; *Faure and Taylor*, 1985].

The ablation rate of the ice field west of the Allan Hills was determined by *Cassidy and Annexstad* [1981] using stakes of a triangulation survey designed to determine the direction and velocity of ice flow in this area [*Annexstad and Schultz*, 1982]. The results from a 2-year study of 22 sites yielded average annual ablation rates of 1.9 ± 0.3 cm yr^{-1} for firn and 4.2 ± 0.2 cm yr^{-1} for exposed ice. (The errors are 1 standard deviation of the mean.) The same sites were subsequently used by *Fudali* [1982] for a gravity survey to determine the thickness of the ice sheet adjacent to the Allan Hills.

This paper reports the results of an extensive ablation study in the Elephant Moraine, the Reckling Moraine, and in the Main ice field west of the Allan Hills. The measurements were made at 265 sites both within and outside of the moraines in order to include

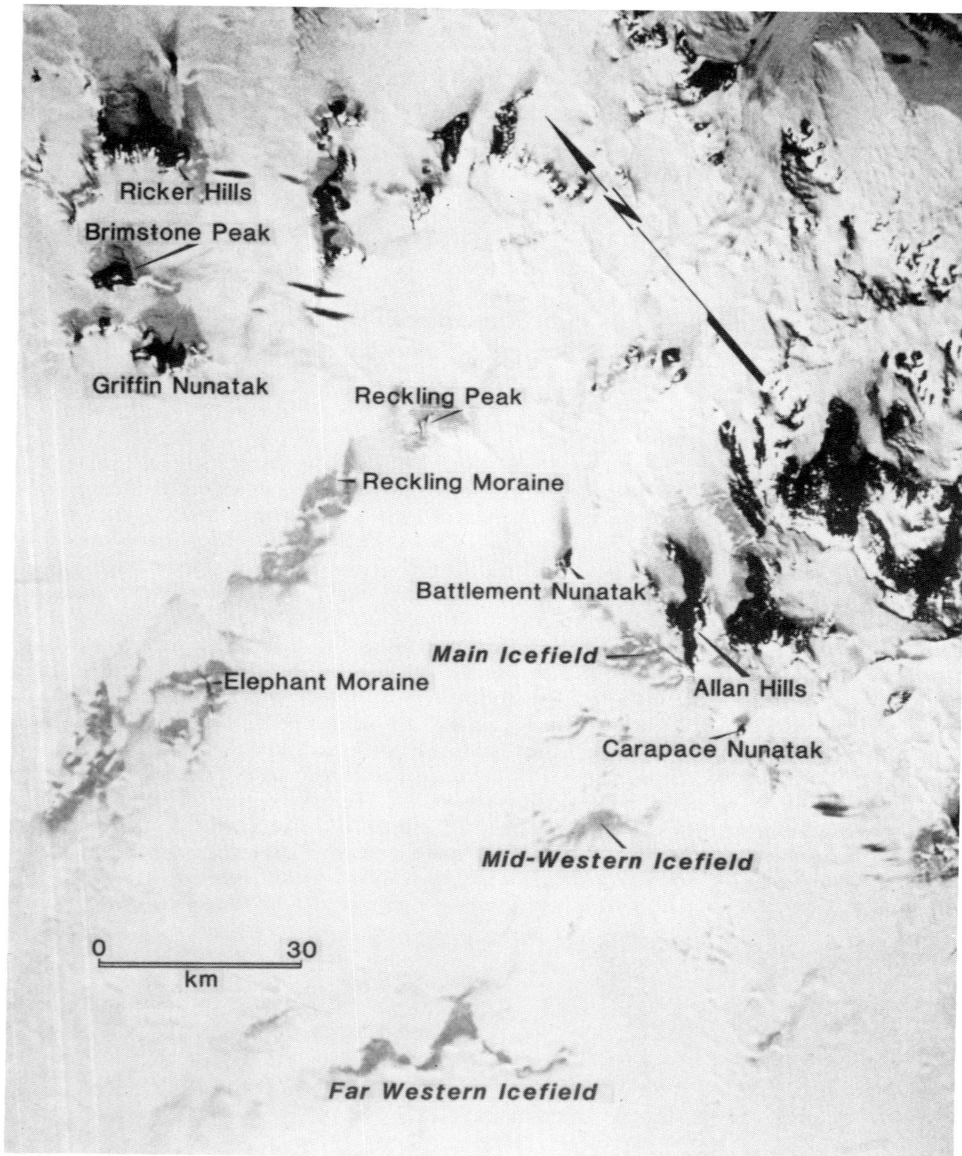

Fig. 1. Landsat photograph of the ice fields west of the Allan Hills showing the locations of the Elephant and Reckling supraglacial moraines.

a wide range of surface conditions and to evaluate the effects of snow and rock clasts on the ablation rates.

METHODS

The ablation rates were measured by means of wooden dowel rods placed in the ice at regular intervals along surveyed lines. Distances within the moraines were measured with a 100-m survey tape. In the ice fields outside the moraines, stations were spaced along surveyed lines by using snowmobile odometers calibrated in miles (Elephant Moraine) or kilometers (Reckling Moraine and Allan Hills), respectively. The dowel rods were placed into holes 10 to 15 cm deep and 1.0 cm in diameter drilled into the ice with an ice screw used for ice climbing. The level of the ice was recorded by cutting a notch into the rod at the level of the ice surface. The lowering of the ice surface at each site was recorded by cutting a second notch into the dowel at the time each station was revisited. The amount of ablation that occurred was measured as the distance between notches using dial calipers calibrated to within 0.001 in. (2.54×10^{-5} m).

This procedure was adopted in order to minimize environmental changes in the ice surface caused by the presence of the wooden dowel rods and to evaluate local microclimatic effects on the ablation rate of the ice sheet. For this reason, all sites were classified with respect to location (moraine or ice field), snow cover, presence of embedded sediment, and type of material (ice or firn) (see appendix Table A1). The presence of the dowel rods did not alter surface conditions at any of the sites either by trapping snow or by increasing or decreasing the ablation rate immediately adjacent to the dowels.

The ablation rates at each of the three areas were recorded both at the end of the field season and 2 years later, thereby providing measurements of both summer and annual rates.

Elephant Moraine

The Elephant Moraine consists of a thin layer of boulders and local accumulations of sand, silt, and clay on bare ice. It is about 5.5 km in length and up to about 1.5 km in width. The rock clasts and sediment are presently accumulating on the surface of the ice sheet as a result of ablation of basal ice, which is exposed here because the flow of the East Antarctic ice sheet is locally disturbed by subglacial bedrock topography [*Faure and Taylor*, 1985; *Faure*, 1990].

The ablation study in and around the Elephant Moraine began in early December 1984 along survey lines shown in Figure 2. The ablation stakes were examined in January 1985 and again in January 1987 when 112 of 155 dowels were recovered. The other 42 dowels were lost because the stations could not be found or because the dowels were buried in snow drifts, had been blown out of their holes by the wind, or had ablated out. In several such cases the dowel rods were found loose in the vicinity of the station where they had been implanted. However, loose dowels were classified as lost and were thus excluded from this study.

The average annual ablation rates observed within the moraine and on the surrounding ice fields, given in Table 1, suggest the following conclusions.

1. The ice underlying the Elephant Moraine ablated more slowly than the surrounding ice fields during the period of this study.

2. The lower ablation rate of ice within the moraine was caused both by the presence of clasts or sediment on the ice and by periodic snow covers.

3. Firn ablated more slowly than ice exposed in the ice fields surrounding the Elephant Moraine.

4. Dirty ice (containing sediment) ablated slightly faster than clean ice within the moraine, but the rate was decreased by the presence of a snow cover.

5. The observed annual ablation rates at similar sites in and around the Elephant Moraine commonly varied by up to 150% of the average values.

The results indicate that ablation rates in and around the Elephant Moraine may have been decreased selectively because of the deposition and migration of snow. However, even the presence of clasts on a snowless ice surface retarded ablation, presumably because the clasts decrease the flow of air across the ice surface and permit the trapped air to approach saturation with water vapor. The slow ablation rate of firn, reported previously by *Annexstad and Schultz* [1982] and confirmed in this study, also may have been caused by the entrapment of air having a higher relative humidity than the ambient air; or the firn may have had a higher albedo than bare ice, thereby reducing solar heating.

Because the ablation rates at some of the stations may have been lowered by the presence of snow, the average annual ablation rates for the stations on each of the four lines extending into the ice fields (labeled A, B, C, and D in Figure 2) were recalculated by omitting measurements that differed from the average of all stations on each line by more than 3 standard deviations of the mean. Lines A and B (see Figure 2) yielded only one measurement each because the other stakes were lost, as discussed above. The resulting weighted average ablation rate of the exposed ice fields surrounding the Elephant Moraine is 4.2 ± 0.2 cm yr^{-1}.

Reckling Moraine

The array of ablation stakes in and around the Reckling Moraine included 77 sites set up in December 1986, of which 63 were recovered in December 1988 (Figure 3). The remainder were lost for the reasons given in the case of the Elephant Moraine. The average ablation rate at 12 stations within the moraine during this period was 1.6 ± 0.3 cm yr^{-1}, whereas 51 sites on the surrounding ice fields ablated at an average rate of 4.1 ± 0.2 cm yr^{-1}. These results are consistent with the average annual ablation rates observed in and around the Elephant Moraine. The average selected ablation rates on the surrounding ice fields (Table 2) range from 4.0 ± 0.2 cm yr^{-1} to 5.6 ± 0.3 cm yr^{-1} and yield a weighted mean of 4.7 ± 0.2 cm yr^{-1}.

The results derived from the Reckling Moraine illustrate how the topographic relief of the ice sheet and the presence of the supraglacial moraine affect the annual ablation rates. Figures 4*a* and 4*b* display the

TABLE 1. Average Annual Ablation Rates (1985–1987) for Different Kinds of Sites in and Around the Elephant Moraine

Description of Site	Ablation Rate, cm yr^{-1} Moraine	Ice Field
Ice, snow-free	2.6 ± 0.1*	4.1 ± 0.1
	0.6 to 4.5, 42†	2.9 to 5.3, 20
Ice, snow-covered	2.2 ± 0.2	3.5
	0.5 to 3.9, 31	1 only
Firn	No sites	2.9 ± 0.3
		2.4 to 3.7, 5
"Dirty" ice, no snow	2.7 ± 0.5	No sites
	0.4 to 4.5, 6	
Dirty ice, snow cover	1.2 ± 0.2	No sites
	0.2 to 2.3, 7	

*All errors are 1 standard deviation of the mean.
†Range of values and number of sites.

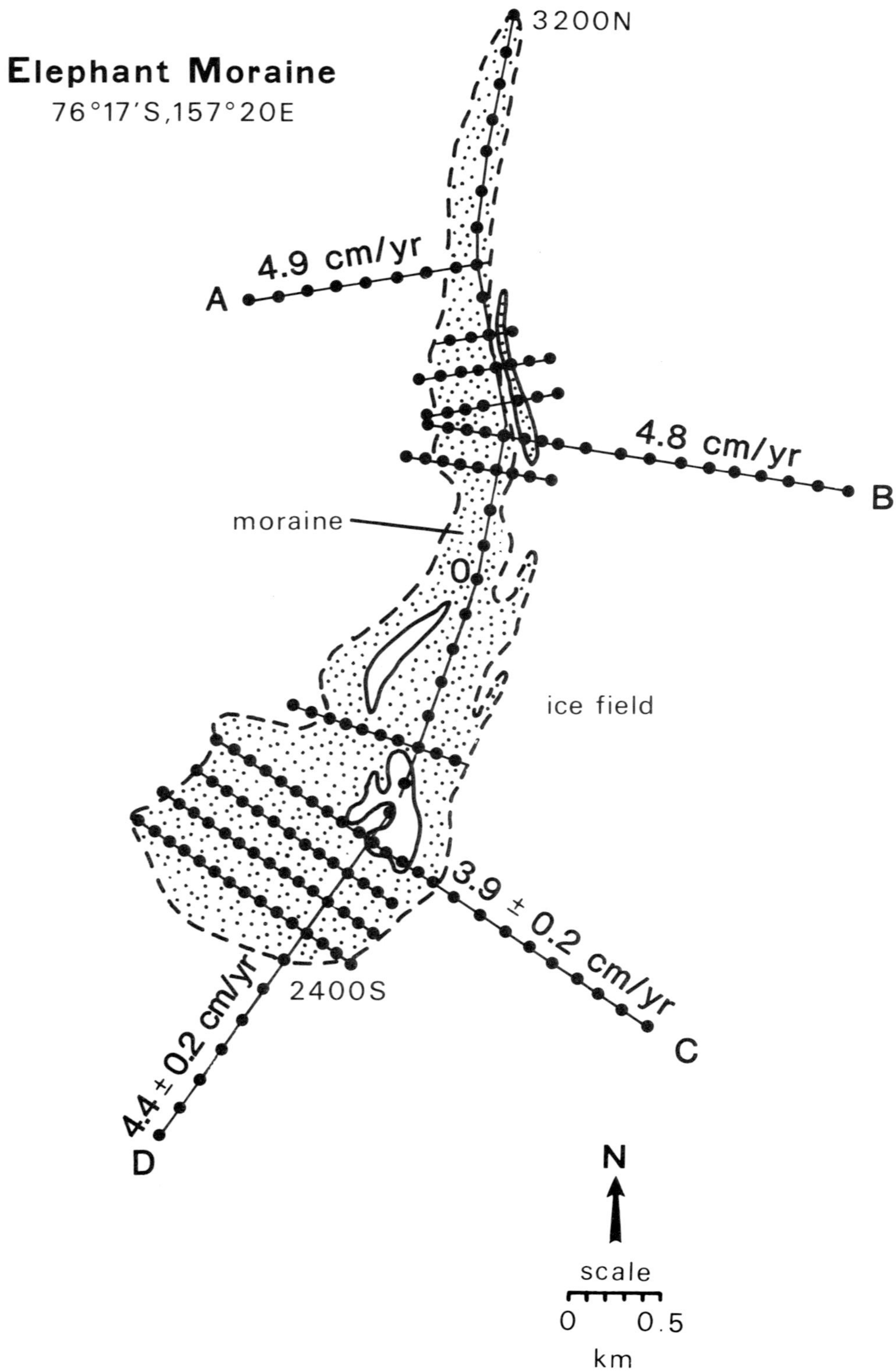

Fig. 2. Survey lines in the Elephant Moraine (see Figure 1) showing locations of ablation stakes and average annual ablation rates.

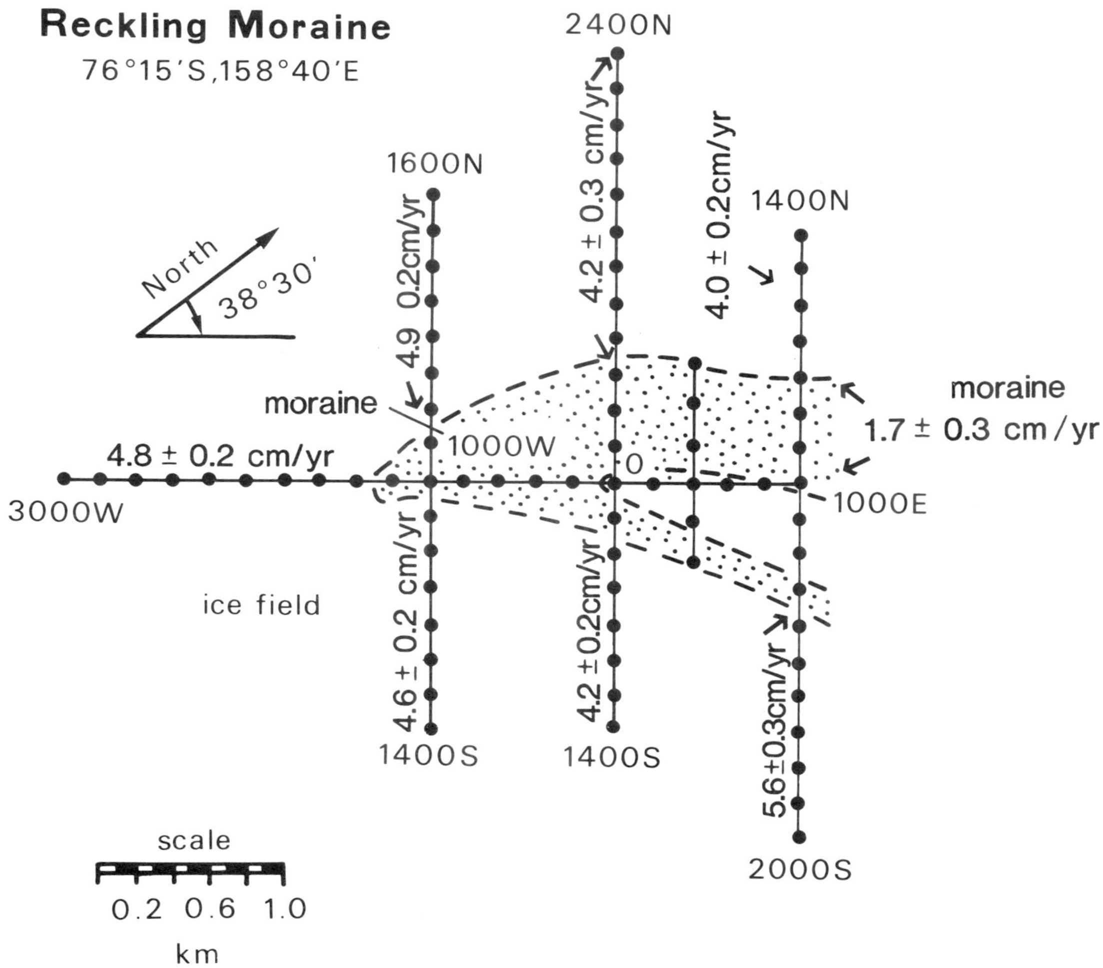

Fig. 3. Survey lines in the Reckling Moraine (see Figure 1) showing locations of ablation stakes and average annual ablation rates.

TABLE 2. Average Annual Ablation Rates on the Ice Fields Surrounding the Reckling Moraine

Survey Lines (Figure 3), m	Average Selected Ablation Rate, cm yr^{-2}
Baseline, 1600 W to 3200 W	4.8 ± 0.2* (6)
Crossline, 1000 W, 600 N to 1600 N	4.9 ± 0.2 (5)
Crossline, 1000 W, 200 S to 1200 S	4.6 ± 0.2 (5)
Crossline, 0, 800 N to 1600 N	4.2 ± 0.3 (4)
Crossline, 0, 600 S to 1500 N	4.2 ± 0.3 (6)
Crossline, 1000 E, 800 N to 1200 N	4.0 ± 0.2 (3)
Crossline, 1000 E, 800 S to 1800 S	5.6 ± 0.3 (5)
Weighted average	4.7 ± 0.2

*All errors are 1 standard deviation of the mean.

ablation rates and topography, respectively, along the crossline at 1000 W from 1400 S to 1600 N, identified in Figure 3. Figure 4*a* clearly shows the decrease in the ablation rate caused by the presence of the moraine. However, the topographic relief along this crossline does not affect the average annual ablation rates of the ice at this locality. Therefore, the topographic relief in the vicinity of the Reckling Moraine is not caused by the ablation of ice.

Allan Hills

The site selected for the ablation study is located at the margin of the East Antarctic ice sheet along the western side of the Allan Hills. The principal survey line (Figure 5) extended from the immediate vicinity of a large rock outcrop west onto the Main ice field. The average annual ablation rate along this line is 4.6 ± 0.3 cm yr^{-1}. The lines extending approximately

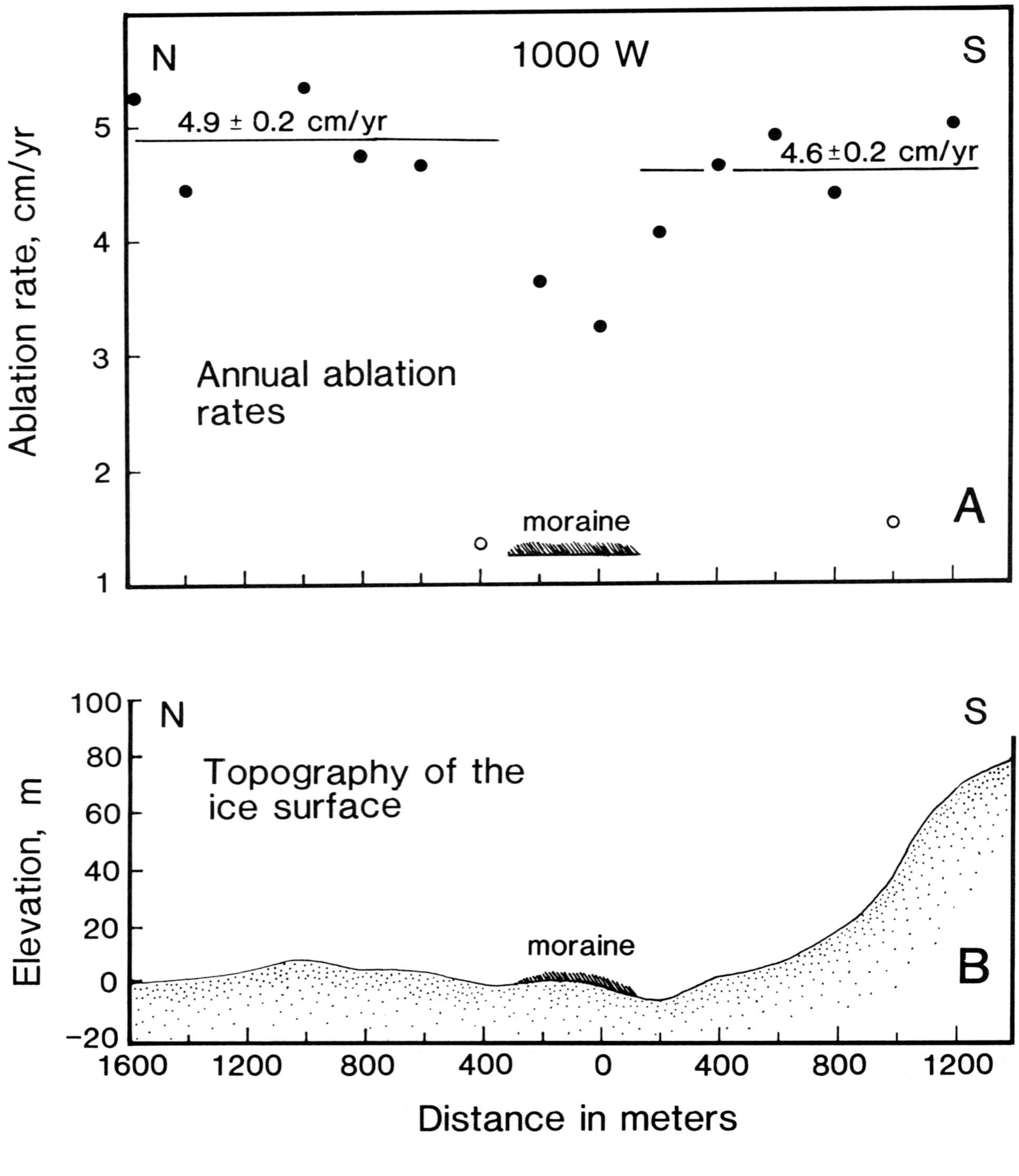

Fig. 4. (*a*) Variation of ablation rates along the crossline at 1000 W (see Figure 3) in the Reckling Moraine. Two values represented by open circles were excluded from the averages because they differ from the means by more than 15 standard deviations. The low ablation rates were most likely caused by snow during part or most of the duration of this study. (*b*) Topographic profile along the crossline at 1000 W in the Reckling Moraine. A comparison of Figures 4*a* and 4*b* suggests that the ablation rate is not affected by the topography of the ice surface but that the presence of the moraine significantly lowers the rate of ablation.

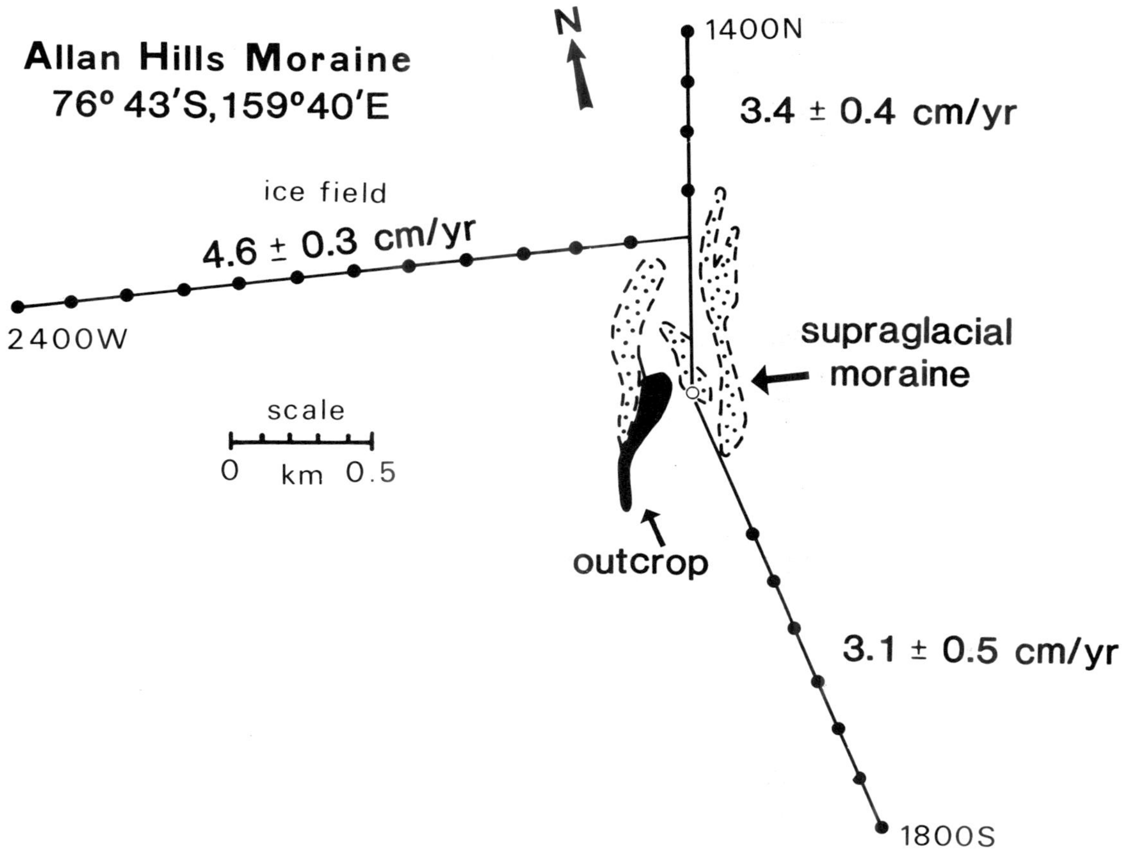

Fig. 5. Survey lines on the Main ice field west of the Allan Hills (see Figure 1) and average annual ablation rates.

north and south from the rock outcrop recorded a lower ablation rate of 3.4 ± 0.3 cm yr^{-1}. These two lines were located within an elongated ice basin that separates the main body of the ice sheet to the west from a local ice tongue on the east. The relatively low ablation rates observed in this basin are probably the result of periodic snow accumulation and are not characteristic of the ice fields near the Allan Hills.

The ablation rates recorded along the main survey line increase from west to east toward the margin of the ice sheet (Figures 6*a* and 6*b*). The values range from about 3.5 cm yr^{-1} at the western end of the line to 5.3 cm yr^{-1} near the margin of the ice sheet and therefore vary inversely as the elevation of the ice surface. These results suggest that the topography of the present surface at this locality is at least partly controlled by the ablation rate, in contrast to the situation in the Reckling Moraine.

Regional Variation of Annual Ablation Rates

The ablation rates of the snow-free ice fields are controlled by climatic factors including the temperature of the ice, the temperature and relative humidity of the air, and the duration and velocity of katabatic winds. The climatic factors at each of the three study sites are influenced to some extent by the elevation above sea level and hence by the distance from the ice edge. Unfortunately, neither the climatic nor the geographic factors of the three sites are well constrained. However, the observed average annual ablation rates appear to increase with decreasing distance from the ice edge.

In the absence of the necessary specific information, the observed annual ablation rates were plotted versus distance west of the longitude of the ice edge at the Allan Hills at about 159°25'E. The resulting diagram (Figure 7) illustrates the implied regional increase of the annual and summer ablation rates toward the eastern edge of the East Antarctic ice sheet between latitudes 76° and 77°S.

Summer Ablation Rates

The data derived from this study permit comparisons between the annual ablation rates and the summertime (December through January) rates at the same sites. The average summer ablation rate

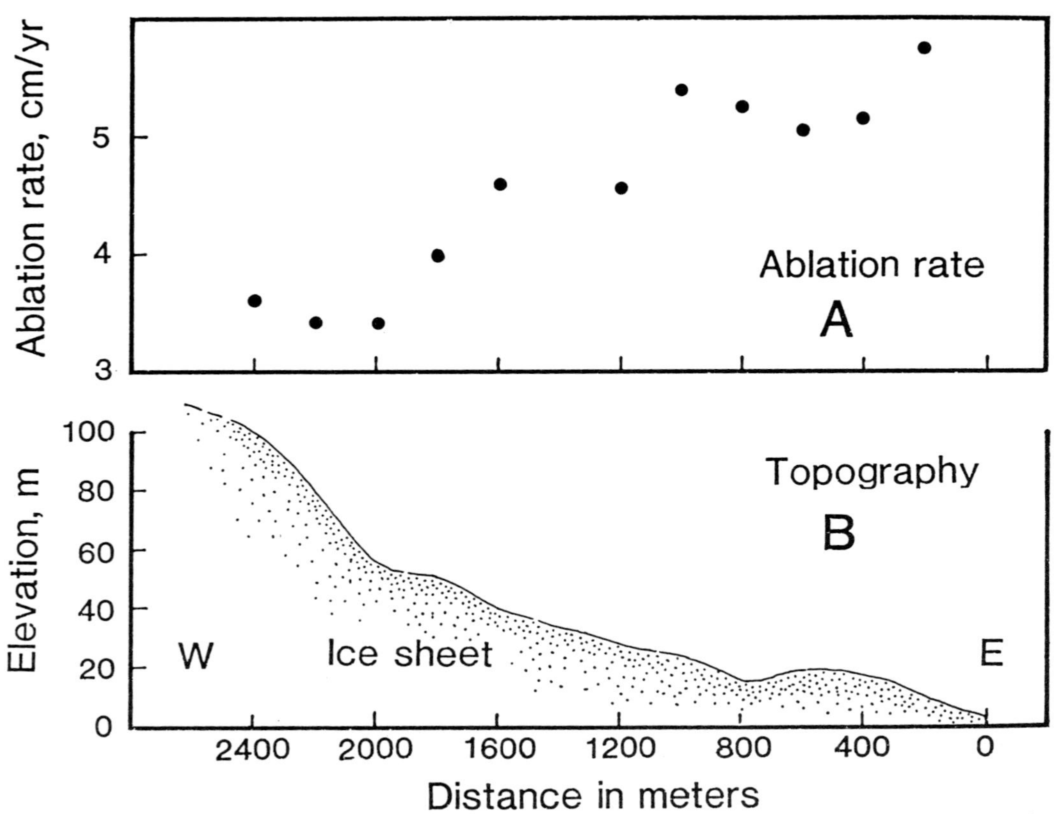

Fig. 6. (*a*) Variation of annual ablation rates along the principal survey line (see Figure 5) at the edge of the East Antarctic ice sheet west of the Allan Hills. (*b*) Topographic profile of the edge of the ice sheet along the principal survey line. A comparison of Figures 6*a* and 6*b* suggests an inverse relation between the ablation rate and the elevation of the ice surface.

TABLE 3. Average Summertime Ablation Rates on the Ice Fields and Moraines West of the Allan Hills

Description of Site	Ablation Rate, mm per 30 days: Moraine	Ice Field
	Elephant Moraine	
Ice, snow-free	13.0 ± 0.9*	16.8 ± 0.9
	1.0 to 20.7, 33†	6.9 to 24.8, 27
Ice, snow-covered	10.6 ± 3.5	No sites
	4.4 to 17.1, 4	
Firn	No sites	12.7 ± 3.3
		9.4 to 16.0, 2
"Dirty" ice,		
snow-free	16.2 ± 4.1	No sites
	6.8 to 45.8, 9	
	Reckling Moraine	
Ice, snow-free	15.3 ± 0.8	18.1 ± 0.7
	11.3 to 17.1, 7	3.2 to 29.1, 48
Ice, snow-covered	4.2 ± 0.9	5
	2.1 to 7.3, 6	One site
	Allan Hills	
Ice, snow-free	No sites	20.9 ± 0.7
		17.1 to 27.9, 20

*All errors are 1 standard deviation of the mean.
†Ranges of values and number of sites.

(1984–1985) within the Elephant Moraine, expressed in millimeters per 30 days in Table 3, was less than that on the surrounding ice fields; the presence of snow tended to reduce the rate further. Firn ablated somewhat more slowly than bare ice, whereas "dirty" ice tended to ablate faster than clean ice in the moraine. Evidently, the summer ablation rates at different kinds of sites in the Elephant Moraine varied much like the annual rates in Table 1. The summertime ablation rates are, however, significantly higher than the annual rates when both are expressed in the same units.

Comparisons at 33 sites within the Elephant Moraine indicate that the ice ablated 8.1 ± 0.9 times faster during the summer than the annual average. However, at eight sites on the ice field the summer ablation rate in 1984–1985 was only 4.5 ± 0.5 times the annual rate at those sites. The greater increase of the summer ablation rate within the moraine may have been caused by the presence of rock clasts which stored solar heat and thereby increased the temperature of the air above the ice surface.

Another way to express the summer ablation rates is to calculate the number of days required for enough ice to ablate at the summer rate to equal the annual

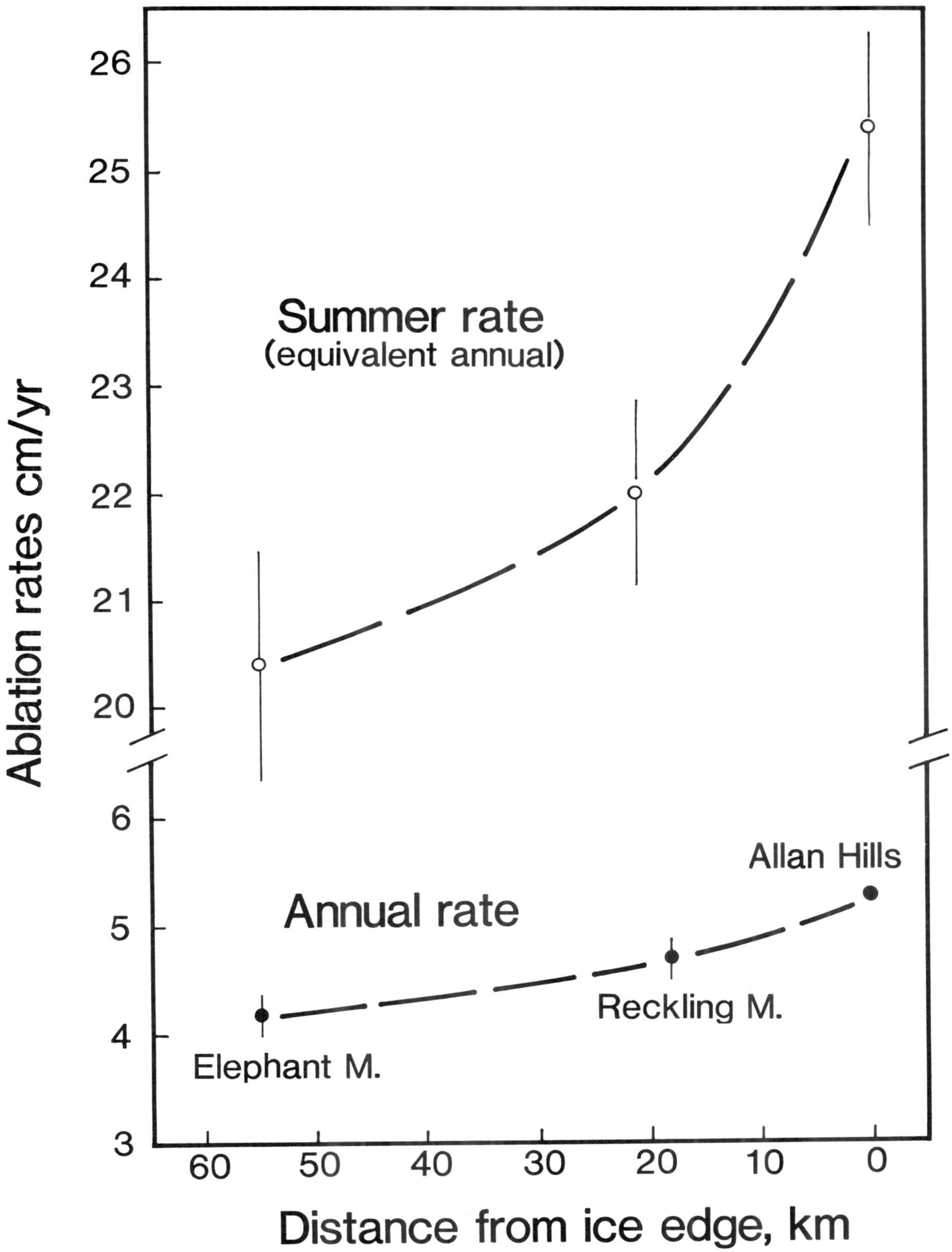

Fig. 7. Regional variation of annual and summer ablation rates of the ice fields adjacent to the study sites. Note that the summer ablation rate has been recalculated to the equivalent annual rate expressed in centimeters per year in order to facilitate the comparison.

loss. This calculation implies that the ablation rate in the winter is assumed to be negligible. At 33 moraine sites this period is 72 ± 11 days, whereas at eight ice field sites the period is 78 ± 7 days. These results imply that the summer ablation rates in and around the Elephant Moraine are in effect for about 10 weeks or less, depending on the ablation rate during the winter months.

The summer ablation rates of snow-free ice in and around the Reckling Moraine (Table 3) were somewhat higher than those of the Elephant Moraine. The highest summer ablation rate occurred on the Main ice field west of the Allan Hills, consistent with the trend displayed by the average annual rates (Figure 7). At 13 sites within the Reckling Moraine the summer ablation rates were 7.3 ± 1.0 times greater than the annual rates at the same stations. At 42 ice field sites near the Reckling Moraine the summertime rates differed from the annual rates only by a factor of 5.1 ± 0.2. The difference was consistent with the results from the Elephant Moraine.

The time required during the summer to ablate a layer of ice equal to the annual amount within the Reckling Moraine is 55 ± 9 days based on 11 moraine sites and 74 ± 3 days at 42 ice field sites. On the ice field adjacent to the Allan Hills, it takes only 61 ± 5 days of ablation at summer rates to equal the annual amount.

The trend in the data suggests that average summer temperatures on the ice fields increase from the Elephant Moraine toward the margin of the ice sheet. As a result, summer ablation rates (see Table 3 and Figure 7) rise toward the margin of the ice sheet, requiring slightly fewer days in the Allan Hills to equal the amount of ice lost annually at each locality.

CONCLUSION

The principal conclusion of this study is that the annual, as well as the summertime, ablation rates increase toward the margin of the ice sheet and that the summer rates are significantly faster than the annual rates. As a result, the total annual thickness of ice ablated at the annual rate could be removed in about 10 weeks at the summer rates, assuming that ablation during the winter is negligible.

The presence of snow and/or rock clasts on the surface of the ice significantly retards ablation, and firn ablates more slowly than solid ice. Both phenomena may be caused by the presence of trapped air having a higher relative humidity than the ambient air over the East Antarctic ice sheet. However, the slower ablation rate of firn can also be attributed to a higher albedo.

The topography of the ice sheet near the Reckling Moraine is independent of the ablation rate and reflects the presence of subglacial bedrock ridges. However, the elevation of the margin of the ice sheet at the Allan Hills appears to be controlled by ablation rather than ice dynamics.

TABLE A1. Classification of Sites and Ablation Rates

Location, m	Ablation Rate: Summer, mm 30 d^{-1}	Code	Annual, cm yr^{-1}	Code
	Elephant Moraine (Figure 2)			
0	45.79	A4	4.47	A4
200 S	9.60	A1	3.81	A1
400 S	19.50	A1	2.64	A1
600 S	11.92	A1	lost	
800 S	4.41	A2	1.63	A1
1000 S	12.85	A1	3.06	A1
1200 S	6.85	B1	lost	
1400 S	6.78	A4	0.43	A4
1600 S	9.42	B3	2.41	B3
1800 S	8.72	A4	0.63	A5
2000 S	4.50	A2	0.47	A2
2200 S	lost		2.26	A2
2400 S	8.21	B1	4.06	B1
2600 S	lost		2.89	B1
2800 S	16.23	B1	lost	
3000 S	15.92	B1	5.06	B1
3200 S	24.76	B1	4.43	B1
3400 S	11.04	B1	4.12	B1
3600 S	14.32	B1	4.43	B1
200 N	n.d.*		n.d.	
400 N	n.d.		n.d.	
600 N	17.70	A1	1.90	A1
800 N	15.60	A1	1.38	A1
1000 N	lost		2.23	A1
1200 N	12.90	B1	lost	
1400 N	n.d.		lost	
1600 N	n.d.		4.48	
1800 N	n.d.		2.48	A1
2000 N	n.d.		4.34	A1
2200 N	n.d.		4.09	A1
2400 N	n.d.		3.09	A1
2600 N	n.d.		lost	
2800 N	n.d.		1.92	A2
3000 N	n.d.		0.81	A1
3200 N	n.d.		2.97	B3
600 N, 100 W	15.44	A1	lost	
600 N, 200 W	9.61	A1	2.39	A2
600 N, 300 W	16.29	A1	2.41	A1
600 N, 400 W	23.15	B1	lost	
600 N, 75 E	11.39	A4	1.58	A5
600 N, 200 E	9.28	B1	lost	
600 N, 300 E	n.d.		3.30	B1
800 N, 100 W	20.65	A1	lost	
800 N, 200 W	11.06	A1	lost	
800 N, 300 W	19.07	A1	lost	
800 N, 400 W	8.32	A1	0.60	A1
800 N, 50 E	lost	(two stakes)	3.22, 2.33	A1, A4
800 N, 100 E	9.65	A1	2.99	A2
800 N, 200 E	17.83	B1	3.72	B1
800 N, 300 E	17.72	B1	3.34	B1
1000 N, 100 W	14.22	A1	lost	
1000 N, 200 W	7.46	A4	3.30	A4
1000 N, 300 W	17.06	A1	lost	
1000 N, 350 W	16.25	A1	2.43	A2
1000 N, 100 E	13.76	A1	lost	
1000 N, 200 E	5.13	A2	2.31	A1
1000 N, 300 E	13.30	A1	lost	
1200 N, 100 W	18.91	A1	4.10	A1
1200 N, 200 W	17.52	A1	lost	
1200 N, 300 W	14.82	A1	3.61	A1

TABLE A1. (continued)

Location, m	Ablation Rate Summer, mm 30 d^{-1}	Code	Annual, cm yr^{-1}	Code
1200 N, 400 W	14.47	A1	lost	
1200 N, 100 E	12.63	A1	n.d.	
1200 N, 200 E	15.62	A1	4.18	A1
1200 N, 300 E	20.95	A1	lost	
1400 N, 100 W	n.d.		2.68	A4
1400 N, 200 W	n.d.		3.52	A1
1400 N, 100 E	n.d.		lost	
1800 N, 300 W	n.d.		5.25	B1
1000 S, 100 W	11.25	A1	1.76	A1
1000 S, 200 W	n.d.		0.24	A5
1000 S, 300 W	17.52	A4	3.19	A4
1000 S, 400 W	4.84	A1	3.03	A2
1000 S, 500 W	23.48	A4	1.44	A5
1000 S, 600 W	16.00	B3	3.14	B3
1000 S, 700 W	21.89	B1	lost	
1000 S, 100 E	17.35	A1	3.51	A2
1000 S, 200 E	18.99	A1	2.48	A1
1000 S, 300 E	12.54	A1	2.04	A2
1600 S, 100 W	14.39	A4	lost	
1600 S, 200 W	9.90	A4	1.93	A5
1600 S, 300 W	n.d.		0.53	A2
1600 S, 400 W	13.85	A1	2.83	A1
1600 S, 500 W	lost		1.76	A1
1600 S, 600 W	12.19	A1	3.43	A2
1600 S, 700 W	10.94	A1	2.15	A1
1600 S, 800 W	lost		0.58	A2
1600 S, 900 W	2.21	A1	2.47	A1
1600 S,1000 W	11.56	B1	lost	
1600 S, 100 E	1.03	A1	lost	
1600 S, 200 E	12.46	A1	lost	
1600 S, 300 E	lost		0.77	A2
1600 S, 400 E	n.d.		3.70	A1
1800 S, 100 W	n.d.		lost	
1800 S, 200 W	n.d.		1.48	A5
1800 S, 300 W	n.d.		1.34	A5
1800 S, 400 W	n.d.		3.77	A1
1800 S, 500 W	n.d.		2.59	A2
1800 S, 600 W	n.d.		3.56	A2
1800 S, 700 W	n.d.		3.89	A2
1800 S, 800 W	n.d.		2.35	A2
1800 S, 900 W	n.d.		lost	
1800 S,1000 W	n.d.		2.98	A1
1800 S, 100 E	n.d.		1.86	A2
1800 S, 200 E	n.d.		3.39	A1
1800 S, 300 E	n.d.		2.40	A2
2000 S, 100 W	n.d.		2.10	A2
2000 S, 200 W	n.d.		2.48	A2
2000 S, 300 W	n.d.		2.54	A2
2000 S, 400 W	n.d.		1.97	A1
2000 S, 500 W	n.d.		1.72	A1
2000 S, 600 W	n.d.		2.82	A2
2000 S, 700 W	n.d.		lost	
2000 S, 800 W	n.d.		lost	
2000 S, 900 W	n.d.		3.97	A2
2000 S,1000 W	n.d.		3.12	A1
2000 S,1100 W	n.d.		4.52	B1
2000 S, 100 E	n.d.		1.74	A1
2000 S, 200 E	n.d.		1.73	A1
2000 S, 300 E	n.d.		1.56	A2
2200 S, 100 W	n.d.		2.53	A2
2200 S, 200 W	n.d.		2.48	A2

TABLE A1. (continued)

Location, m	Ablation Rate Summer, mm 30 d^{-1}	Code	Annual, cm yr^{-1}	Code
2200 S, 300 W	n.d.		2.57	A1
2200 S, 400 W	n.d.		2.13	A2
2200 S, 500 W	n.d.		0.96	A2
2200 S, 600 W	n.d.		1.77	A1
2200 S, 700 W	n.d.		1.57	A2
2200 S, 800 W	n.d.		2.44	A2
2200 S, 900 W	n.d.		3.92	B1
2200 S,1000 W	n.d.		2.62	A1
2200 S, 100 E	n.d.		1.09	A1
2200 S, 200 E	n.d.		3.74	A1
2200 S, 300 E	n.d.		2.91	A1
1800 N, 300 W (line A, Fig. 2)				
0.1 mi†			lost	
0.2 mi	n.d.		lost	
0.3 mi	n.d.		4.85	B1
0.4 mi	n.d.		lost	
0.5 mi	n.d.		2.36	B3
0.6 mi	n.d.		3.74	B3
800 N, 300 E (line B, Fig. 2)				
0.1 mi	21.21	B1	lost	
0.2 mi	21.27	B1	lost	
0.3 mi	21.04	B1	4.79	B1
0.4 mi	19.63	B1	lost	
0.5 mi	22.44	B1	lost	
0.6 mi	14.24	B1	lost	
0.7 mi	21.80	B1	lost	
0.8 mi	17.17	B1	lost	
0.9 mi	18.93	B1	lost	
1.0 mi	16.23	B1	lost	
1600 S, 400 E (line C, Fig. 2)				
0.1 mi	n.d.		3.61	B1
0.2 mi	n.d.		3.71	B1
0.3 mi	n.d.		3.45	B2
0.4 mi	n.d.		4.06	B1
0.5 mi	n.d.		4.72	B1
0.6 mi	n.d.		4.33	B1
0.7 mi	n.d.		3.73	B1
0.8 mi	n.d.		3.03	B2
0.9 mi	n.d.		lost	
	Reckling Moraine (Figure 3)			
0	3.66	A2	lost	
200 W	lost		lost	
400 W	13.41	A1	1.64	A2
600 W	lost		0.37	A2
800 W	17.11	A1	1.88	A2
1000 W	15.69	A1	3.24	A2
1200 W	lost		lost	
1400 W	lost		1.62	B1
1600 W	18.70	B1	4.91	B1
1800 W	16.20	B1	4.72	B1
2000 W	22.65	B1	5.36	B1
2200 W	17.71	B1	5.47	B1
2400 W	lost		lost	
2600 W	9.37	B1	4.05	B1
2800 W	19.55	B1	3.37	B1
3000 W	lost		1.31	B1
3200 W	lost		4.50	B1
200 E	lost		4.02	B1
400 E	17.06	B1	4.65	B1
600 E	19.94	B1	3.53	B1

TABLE A1. (continued)

Location, m	Ablation Rate Summer, mm 30 d^{-1}	Code	Annual, cm yr^{-1}	Code
800 E	16.08	B1	3.39	B1
1000 E	16.55	A1	1.84	A1
0, 100 N	17.10	A4	1.98	A5
400 N	lost		lost	
500 N	7.28	A2	lost	
600 N	2.70	A2	1.04	A2
700 N	14.01	B1	2.79	B1
800 N	3.17	B1	3.39	B1
1000 N	18.50	B1	4.82	B1
1200 N	lost		lost	
1400 N	19.42	B1	4.20	B1
1600 N	15.58	B1	4.40	B1
1800 N	lost		2.67	B1
2000 N	lost		lost	
2200 N	lost		lost	
2400 N	lost		lost	
100 S	2.94	A2	0.66	A2
200 S	lost		lost	
300 S	lost		lost	
400 S	5.02	B2	1.69	B2
600 S	15.28	B1	3.63	B1
800 S	24.90	B1	5.16	B1
1000 S	17.28	B1	4.30	B1
1200 S	14.97	B1	3.64	B1
1400 S	13.38	B1	4.81	B1
1500 S	10.25	B1	3.62	B1
400 E, 100 N	lost		lost	
400 E, 200 N	lost		0.89	A1
400 E, 300 N	16.91	A1	1.54	A2
400 E, 400 N	6.70	A2	0.98	A2
400 E, 600 N	lost		lost	
400 E, 700 N	lost		lost	
400 E, 100 S	lost		lost	
400 E, 200 S	15.34	B1	lost	
400 E, 300 S	lost		lost	
400 E, 400 S	2.09	A2	3.53	B1
1000 E, 200 N	9.77	A4	1.78	A2
1000 E, 400 N	lost		lost	
1000 E, 600 N	16.43	B1	lost	
1000 E, 800 N	13.86	B1	3.81	B1
1000 N	19.02	B1	3.72	B1
1200 N	19.25	B1	4.45	B1
1400 N	20.59	B1	lost	
1600 N	18.94	B1	lost	
200 S	13.40	B1	2.94	B1
400 S	16.15	B1	4.14	B1
600 S	lost		4.35	B1
800 S	18.58	B1	4.85	B1
1000 S	25.02	B1	5.57	B1
1200 S	22.31	B1	5.11	B1
1400 S	10.54	B1	2.15	B1
1600 S	29.12	B1	6.85	B1
1800 S	19.61	B1	5.69	B1
2000 S	16.46	B1	lost	
1000 W, 200 N	11.32	A1	3.62	A2
1000 W, 400 N	lost		1.36	B1
1000 W, 600 N	18.12	B1	4.67	B1
1000 W, 800 N	19.92	B1	4.73	B1
1000 W,1000 N	21.22	B1	5.35	B1
1000 W,1200 N	24.11	B1	lost	
1000 W,1400 N	21.71	B1	4.45	B1

TABLE A1. (continued)

Location, m	Ablation Rate Summer, mm 30 d^{-1}	Code	Annual, cm yr^{-1}	Code
1000 W,1600 N	25.71	B1	5.26	B1
1000 W, 200 S	20.67	B1	4.04	B1
1000 W, 400 S	22.88	B1	4.65	B1
1000 W, 600 S	16.13	B1	4.90	B1
1000 W, 800 S	18.13	B1	4.39	B1
1000 W,1000 S	15.29	B1	1.52	B1
1000 W,1200 S	20.02	B1	5.00	B1
1000 W,1400 S	23.28	B1	lost	
Allan Hills (Figure 5)				
1000 N	26.13	B1	lost	
1200 N	20.67	B1	3.02	B1
1400 N	23.25	B1	3.83	B1
600 S	22.09	B1	4.55	B1
800 S	26.52	B1	2.66	B1
1000 S	22.86	B1	2.92	B2
1200 S	lost		lost	
1400 S	23.94	B1	lost	
1600 S	17.81	B1	lost	
1000 S	18.28	B1	lost	
600 N, 200 W	17.36	B1	5.74	B1
600 N, 400 W	18.60	B1	5.16	B1
600 N, 600 W	21.48	B1	5.06	B1
600 N, 800 W	17.08	B1	5.27	B1
600 N, 1000 W	27.91	B1	5.41	B1
600 N, 1200 W	18.44	B1	4.55	B1
600 N, 1400 W	21.42	B1	lost	
600 N, 1600 W	18.22	B1	4.61	B1
600 N, 1800 W	lost		4.00	B1
600 N, 2000 W	18.85	B1	3.42	B1
600 N, 2200 W	17.09	B1	3.44	B1
600 N, 2400 W	19.30	B1	3.63	B1

Site codes: A, moraine site; B, ice field site; 1, snow-free; 2, snow-covered at the time of recovery; 3, firn; 4, "dirty" ice, snow-free; and 5, dirty ice, snow-covered at the time of recovery.

*n.d., not determined.

†1 mile = 1.608 km.

Acknowledgments. We were ably assisted in the field by K. S. Taylor in 1984–1985, by M. L. Strobel and E. H. Hagen in 1986–1987, and by R. J. Elliot and T. A. Aebie in 1988–1989. We thank Kathleen L. Doddroe for typing the manuscript and John Nagy for drafting the diagrams. We also thank two anonymous reviewers for helpful comments. This is contribution 88 of the Laboratory for Isotope Geology and Geochemistry, Ohio State University. This study was supported by National Science Foundation grants DPP-8314136 and DPP-8716088.

REFERENCES

Annexstad, J. O., and L. Schultz, Triangulation survey of the Allan Hills icefield, 1981–1982, *Antarct. J. U.S., 17,* 57–58, 1982.

Cassidy, W. A., and J. O. Annexstad, Antarctic search for meteorites, 1980–1981, *Antarct. J. U.S., 16,* 61–62, 1981.

Cassidy, W. A., E. Olsen, and K. Yanai, Deep-freeze storehouse for meteorites, *Science, 198,* 727–731, 1977.

Faure, G., Subice topography and formation of supraglacial moraines (abstract), *LPI Tech. Rep. 90-03,* 32–33, 1990.

Faure, G., and K. S. Taylor, The geology and origin of the Elephant Moraine on the East Antarctic ice sheet, *Antarct. J. U.S., 20,* 11–12, 1985.

Fudali, R. F., Gravity measurements across the Allan Hills main meteorite collecting area, *Antarct. J. U.S., 17,* 58–60, 1982.

Montello, J. M., The provenance of till at the Elephant Moraine on the East Antarctic ice sheet, M.S. thesis, 170 pp., Dep. of Geol. and Mineral., Ohio State Univ., Columbus, 1986.

Taylor, K. S., Lithologies and distribution of clasts in the Elephant Moraine, Allan Hills, south Victoria Land, Antarctica, M.S. thesis, 121 pp., Dep. of Geol., Kent State Univ., Kent, Ohio, 1986.

Whillans, I. M., and W. A. Cassidy, Catch a falling star: Meteorites and old ice, *Science, 222,* 55–57, 1983.

(Received September 24, 1990;
accepted January 30, 1991.)

CONTRIBUTIONS TO ANTARCTIC RESEARCH II
ANTARCTIC RESEARCH SERIES, VOLUME 53, PAGES 33–65

CONTRIBUTIONS OF SEDIMENTOLOGIC, THERMAL ALTERATION, AND ORGANIC CARBON DATA TO PALEOENVIRONMENTAL INTERPRETATION OF FINE-GRAINED PERMIAN CLASTICS FROM THE BEARDMORE GLACIER REGION, ANTARCTICA

T. C. HORNER AND L. A. KRISSEK

Byrd Polar Research Center and Department of Geological Sciences, Ohio State University, Columbus, Ohio 43210

Field relations and total organic carbon (TOC) contents have been studied in the finer-grained portions of four Permian formations from the central Transantarctic Mountains. Special emphasis has been given to the fine-grained facies because of their potential significance as carriers of paleoclimatic data. In ascending order, the Permian section consists of the Pagoda, Mackellar, Fairchild, and Buckley formations. The depositional environments of these units have been determined on the basis of field relationships, lithologic associations, and sedimentary structures. The Pagoda Formation is composed of sheared gray-green diamictite with minor amounts of sandstone, stiltstone, and shale and is interpreted to be a glacial deposit. The Mackellar Formation is composed of interbedded sandstone, siltstone, and shale, coarsens upward on the scale of the entire formation, and is interpreted to be a shallow deltaic deposit. The Fairchild Formation is dominated by cross-bedded medium- to coarse-grained channel-form sandstone beds and is interpreted to be a braided stream deposit. The Buckley Formation consists of medium- to coarse-grained cross-bedded sandstone with lesser amounts of siltstone, shale, and coal and was formed by braided stream systems and the associated floodplain deposition. The amount of organic matter in these fine-grained sedimentary rocks is controlled both by factors that affected the amount of organic matter originally supplied to the sediments and by factors that have affected the amount of organic matter preserved in the sediments. Lithologic and geochemical data indicate that the grain size, type of organic matter, degree of bioturbation, and dissolved oxygen content did not vary significantly during deposition of these fine-grained Permian clastics. As a result, significant variability in TOC content appears to reflect variations in organic carbon input due to changes in productivity or climate and/or postdepositional destruction of TOC during thermal alteration by Jurassic sills and dikes. Profiles of vitrinite reflectance and TOC contents suggest that samples taken within one sill thickness adjacent to a sill have been significantly altered. Organic carbon contents from unaltered shale and siltstone facies vary between the four formations studied. The fine-grained intervals of the Pagoda, Mackellar, Fairchild, and Buckley formations have mean TOC contents of 0.20%, 0.33%, 0.98%, and 21.64%, respectively. These large-scale (interformational) differences in the mean TOC content of unaltered samples are interpreted to record an upsection increase in regional productivity, consistent with the general pattern of climatic warming and the decay of the Carboniferous/Permian ice sheet. Smaller-scale variations in TOC content are attributed to local differences in depositional environment and the resulting minor fluctuations in grain size and rates of deposition. TOC contents help to distinguish between peat-forming and pond facies in the Buckley Formation but cannot be used alone to define subenvironments within the glacial, deltaic, and fluvial systems. The most comprehensive interpretations of paleoenvironments use all available data, including field observations, compositional information, and TOC contents.

INTRODUCTION

Permian clastics from the Beardmore Glacier region are part of an outcrop belt that extends for more than 1000 km along the paleo-Pacific margin of east Antarctica. Exposures from the Nimrod Glacier to the Pensacola Mountains contain thick upper Paleozoic and Mesozoic sedimentary sequences that overlie a Precambrian and lower Paleozoic basement complex of plutonic, metamorphic, and metavolcanic rocks. The continuity of this succession has been attributed to development along an evolving foreland basin [*Collinson*, 1990]. Initial sedimentary deposition in the basin is indicated by the preservation of Devonian sandstones and conglomerates, which are overlain disconformably by Carboniferous(?)-Permian diamictites, shales, siltstones, and sandstones [*Elliot*, 1975]. Continued subsidence preserved 2500 m of Permian and Triassic rocks in the Beardmore Glacier area; these rocks were subsequently exposed by uplift of the present-day Transantarctic Mountains. The focus of this study is fine-grained clastics from the Permian portion of the Victoria Group, with emphasis on the paleoclimatic and paleoenvironmental information that can be interpreted from these units.

METHODS

Fieldwork for this project was conducted during the 1985–1986 field season from a remote base camp that was established on the Bowden Névé in the Beardmore Glacier region of the central Transantarctic Mountains. Twenty-four sections from 21 localities were described and measured, and 310 samples were collected for chemical and mineralogic analysis [*Horner and Krissek*, 1989; *Krissek and Horner*, 1988*a*, *b*]. These sections total 3789 m in thickness and provide good control of stratigraphic and geographic variations within this part of the Permian depositional basin (Figure 1). In an effort to minimize the chemical variability introduced by differences in grain size, mineralogy, and permeability, samples were collected only from shales and siltstones; and only the finest-grained beds from each measured section were sampled. Fresh samples were quarried from each sample locality to prevent contamination by float and to minimize surficial weathering effects.

In addition to field descriptions, samples were analyzed to determine vitrinite reflectance and carbon content. Vitrinite is a residual carbon compound that becomes reflective when heated. The degree of reflectivity is directly related to the amount of heating, so vitrinite reflectance provides a very accurate method for determining the extent of alteration in sedimentary rocks. Vitrinite reflectance measurements were performed by a commercial laboratory, and carbon content was measured using a coulometric carbon analyzer at the Ohio State University. Total organic carbon (TOC) values (in weight %) were obtained by oxidizing the samples at 900°C to measure total carbon content and then analyzing a split of the same sample to measure the inorganic (carbonate) carbon content. The TOC content is calculated by difference:

$$\text{TOC} = \text{total carbon} - \text{inorganic carbon}$$

Instrument error for all TOC measurements was determined by running duplicate samples for each analysis and running every tenth sample four times. Results indicate very good reproducibility of total carbon and inorganic carbon contents, with maximum (worst case) standard deviations from replicate sample runs equal to 0.02% for the Pagoda Formation, 0.04% for the Mackellar Formation, 0.03% for the Fairchild Formation, and 0.88% for the Buckley Formation. Higher standard deviation in the Buckley Formation accompanies a higher average carbon content and still supports the internal consistency of TOC measurements.

STRATIGRAPHY

The four units studied are, in ascending order, the Pagoda, Mackellar, Fairchild, and Buckley formations. These units range in age from Late Carboniferous(?) to Permian [*Barrett et al.*, 1972; *Kyle and Schopf*, 1982] and form the lower portion of the Victoria Group (Figure 2). The upper portion of the Victoria Group consists of the Triassic Fremouw and Falla formations and the Triassic/Jurassic(?) Prebble Formation. The Victoria Group is underlain by the Devonian Taylor Group and overlain by the Jurassic Ferrar Group. These units combine to form the Beacon Supergroup. Beacon sediments and their lateral equivalents are exposed along the paleo-Pacific margin of Antarctica from northern Victoria Land to the Ellsworth Mountains and may record deposition in an evolving foreland basin [*Collinson*, 1990].

SEDIMENTOLOGY AND ENVIRONMENTS OF DEPOSITION

Our interpretations of environments of deposition and the sedimentary structures and lithologic relationships that support those interpretations are summarized below. Our interpretations generally agree with those proposed by other researchers [*Grindley*, 1963; *Barrett*, 1968; *Lindsay*, 1968; *Barrett et al.*, 1986; *Miller et al.*, 1987; *Miller*, 1989; *Isbell*, 1990]. These previous studies are supplemented by our field observations and our selective laboratory investigation of the fine-grained facies.

Pagoda Formation

The Pagoda Formation is interpreted to be a glacial deposit [*Grindley*, 1963; *Lindsay*, 1968, 1970; *Waugh*, 1988; *Miller*, 1989] that formed during the Carboniferous/Permian glaciation of the Antarctic continent. Lithologies within the Pagoda Formation are highly variable, with most measured sections composed predominantly of massive diamictite and lesser amounts of sandstone, siltstone, and shale. The average composition of the Pagoda Formation (as measured in this study) is shown in Table 1.

Three stratigraphic sections are used here to illustrate the extremes of composition within the Pagoda Formation. Section MMP (Figure 3) is a typical diamictite-rich section. Sedimentary structures and depositional environments of the diamictite have been described by *Lindsay* [1968], *Waugh* [1988], and *Miller* [1989]. Diamictite from the Pagoda Formation is gray-green, poorly sorted, and massive or sheared. The diamictite can be structureless, slumped, or

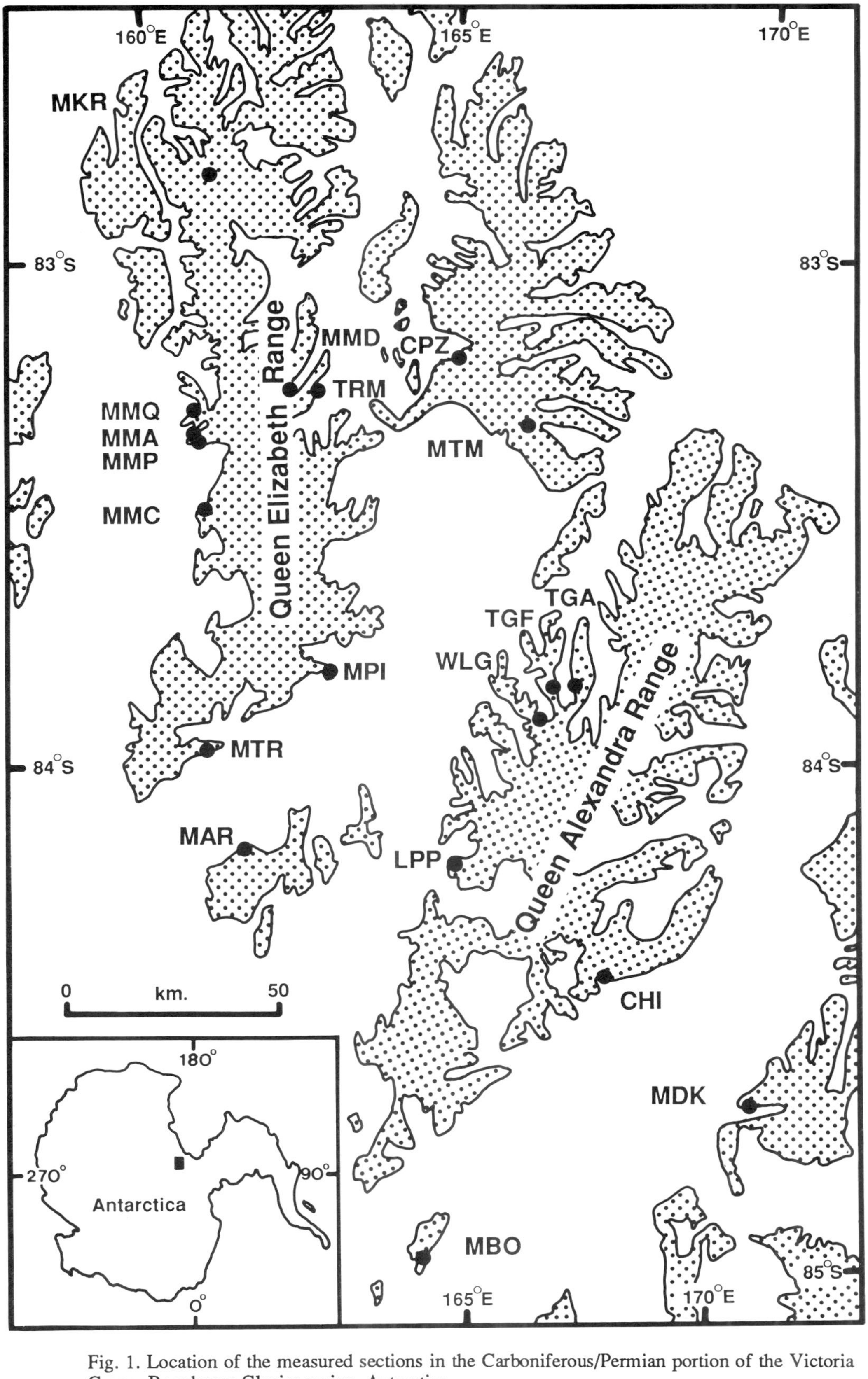

Fig. 1. Location of the measured sections in the Carboniferous/Permian portion of the Victoria Group, Beardmore Glacier region, Antarctica.

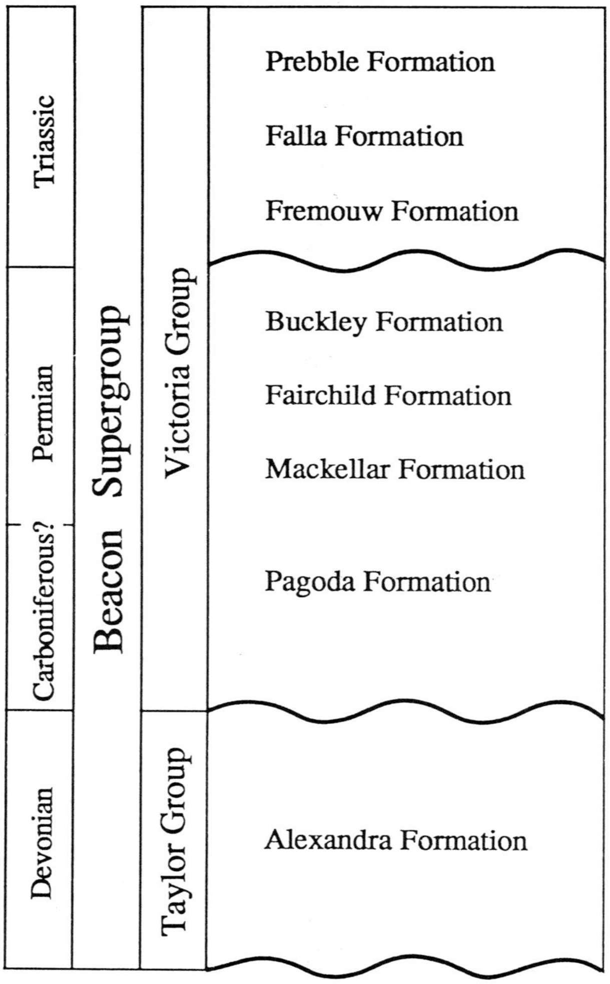

Fig. 2. Stratigraphy of the Victoria Group and associated units, Beardmore Glacier region, Antarctica. Wavy lines indicate unconformities. (After *Barrett et al.* [1986].)

interfingered with sandstone pods and contains clasts of granite, quartzite, chert, and metasediment in a matrix of clay, silt, and poorly sorted sand grains. The sand component is quartz-rich and is interpreted to have been derived from the underlying Alexandra Formation, whereas the granite and metamorphic clasts originated from an unknown source that presently lies beneath the Antarctic ice sheet. The diamictites are interpreted as basal debris (ground moraine) that was deposited by moving ice [*Lindsay*, 1968]. *Miller* [1989] distinguished diamictite deposited by advancing ice from finer-grained units deposited during ice retreat.

Sandstone is less common than diamictite in most Pagoda sections but is the dominant lithology in section CHI (Figure 4). Sandstones are fine to medium grained and form lenticular, pod-shaped, or planar bodies. Sedimentary structures in the sandstone

TABLE 1. Lithologic Summary of the Pagoda Formation, Beardmore Glacier Region, Antarctica

Section	Lithology			
	Shale	Siltstone	Sandstone	Diamictite
MMP	...	1.5 m	4.0 m	69.5 m
MMC	...	...	...	17.6 m
CPZ	36.5 m	43.0 m	...	30.5 m
MTM	20.5 m	10.0 m	10.0 m	74.3 m
TGA	...	4.0 m	22.0 m	113.2 m
CHI	...	14.6	44.8 m	6.0 m
MBO	...	...	...	18.5 m
Total	57.0 m	73.1 m	80.8 m	329.6 m
Percentage	10%	14%	15%	61%

beds include trough cross-bedding, ripple marks, and soft-sediment deformation in the form of contorted bedding. The sediment varies from well sorted to poorly sorted (gradational to diamictite). Sandstone bodies most commonly occur as lenticular forms tens of meters wide, with erosional lower boundaries and internal cross-bedding. Smaller pods of sandstone are very abundant and range from several centimeters to 2 m in width. Sandstone pods interfinger laterally with diamictite or siltstone deposits. Planar sandstone beds are rare and may have been formed by migrating sand waves within proglacial streams. The combination of sedimentary structures and the lithologic association with nearby diamictite indicate that the sandstone bodies were deposited by subglacial or fluvial outwash streams [*Lindsay*, 1968; *Waugh*, 1988; *Miller*, 1989].

The Pagoda Formation at section CPZ (Figure 5) contains abundant fine-grained material. Fine-grained Pagoda sediments occur as shale, silty shale, and poorly sorted sandy mudstone. Shale and silty shale are rare in most sections but form significant proportions of sections CPZ, MTM, and CHI. Siltstones within these sections contain ripple marks, slump structures, flaser bedding, and wavy bedding. These sedimentary structures indicate rapid deposition in a system with abundant sediment. Fine-grained deposits are interpreted to have formed close to the ice margin in deeper ponds or small lakes [*Lindsay*, 1968; *Waugh*, 1988; *Miller*, 1989]. The rarity of dropstones in these quiet water deposits suggests that adjacent land areas did not experience continuous glaciation. Environments where deposition occurred in standing water were probably more common during the glacial retreats than during the glacial advances, since glacial retreats were the time of maximum meltwater production, which enhanced formation of small proglacial lakes. In contrast, glacial advances blanketed much of the study area with ice, and cooler temperatures during these advances reduced the

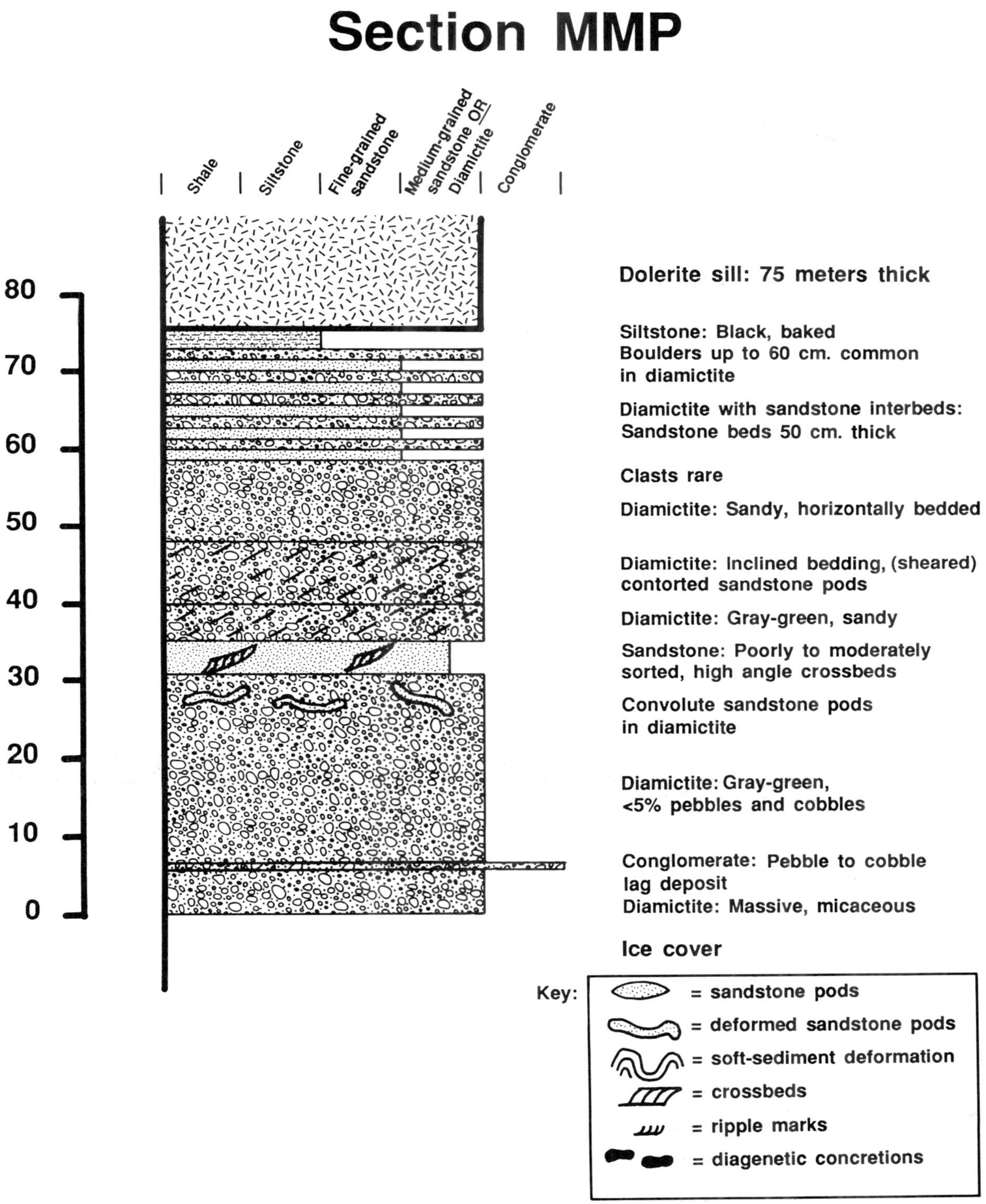

Fig. 3. Diamictite-dominated, measured section MMP from the Pagoda Formation, Moore Mountains. Diamictites are an end-member in the range of glacial deposits and indicate ice advance or retreat.

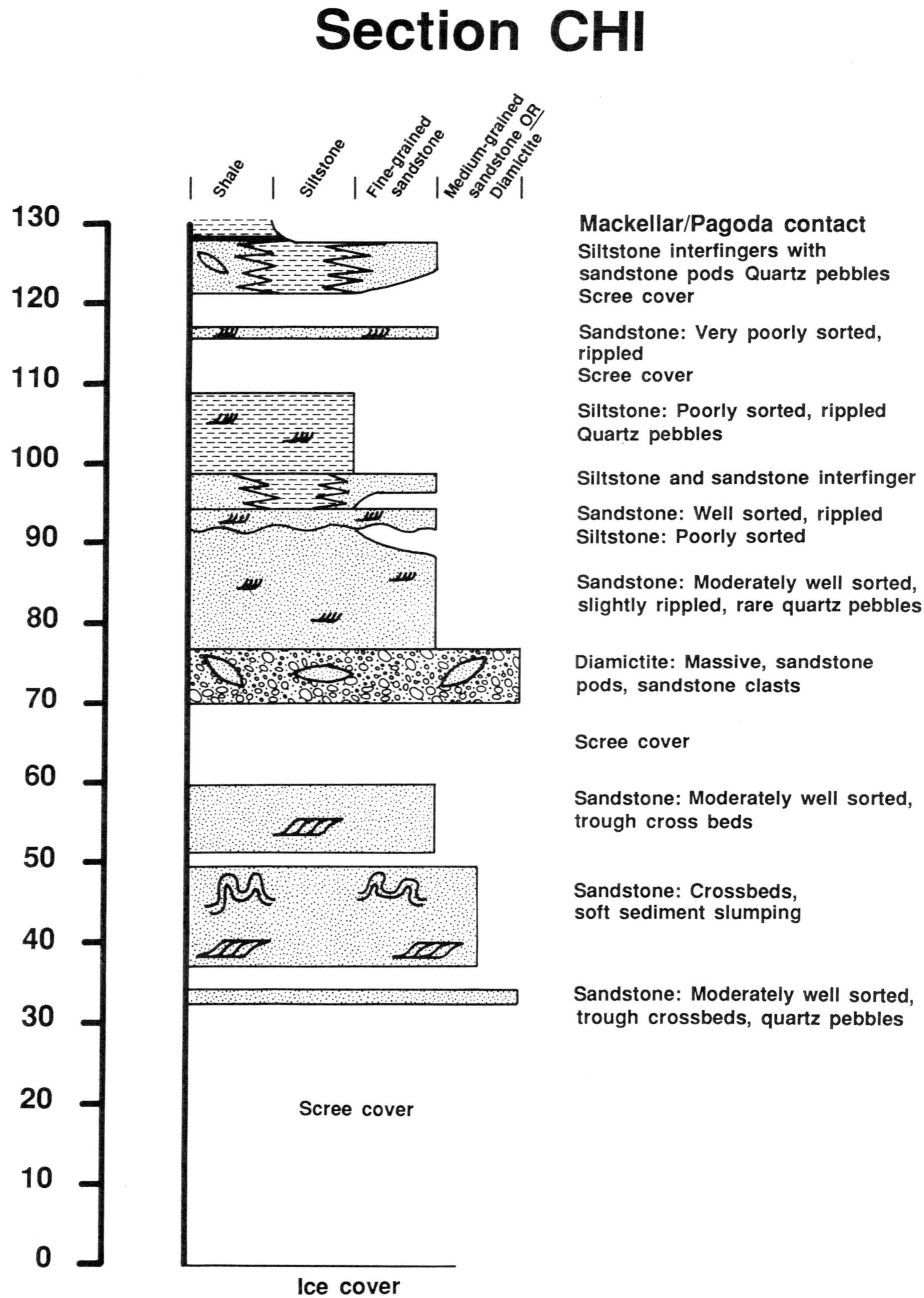

Fig. 4. Sandstone-dominated, measured section CHI from the Pagoda Formation, Cherry Icefall. Sandstones are an end-member in the spectrum of glacial deposits and indicate deposition by moving water. (See Figure 3 for the key to the symbols.)

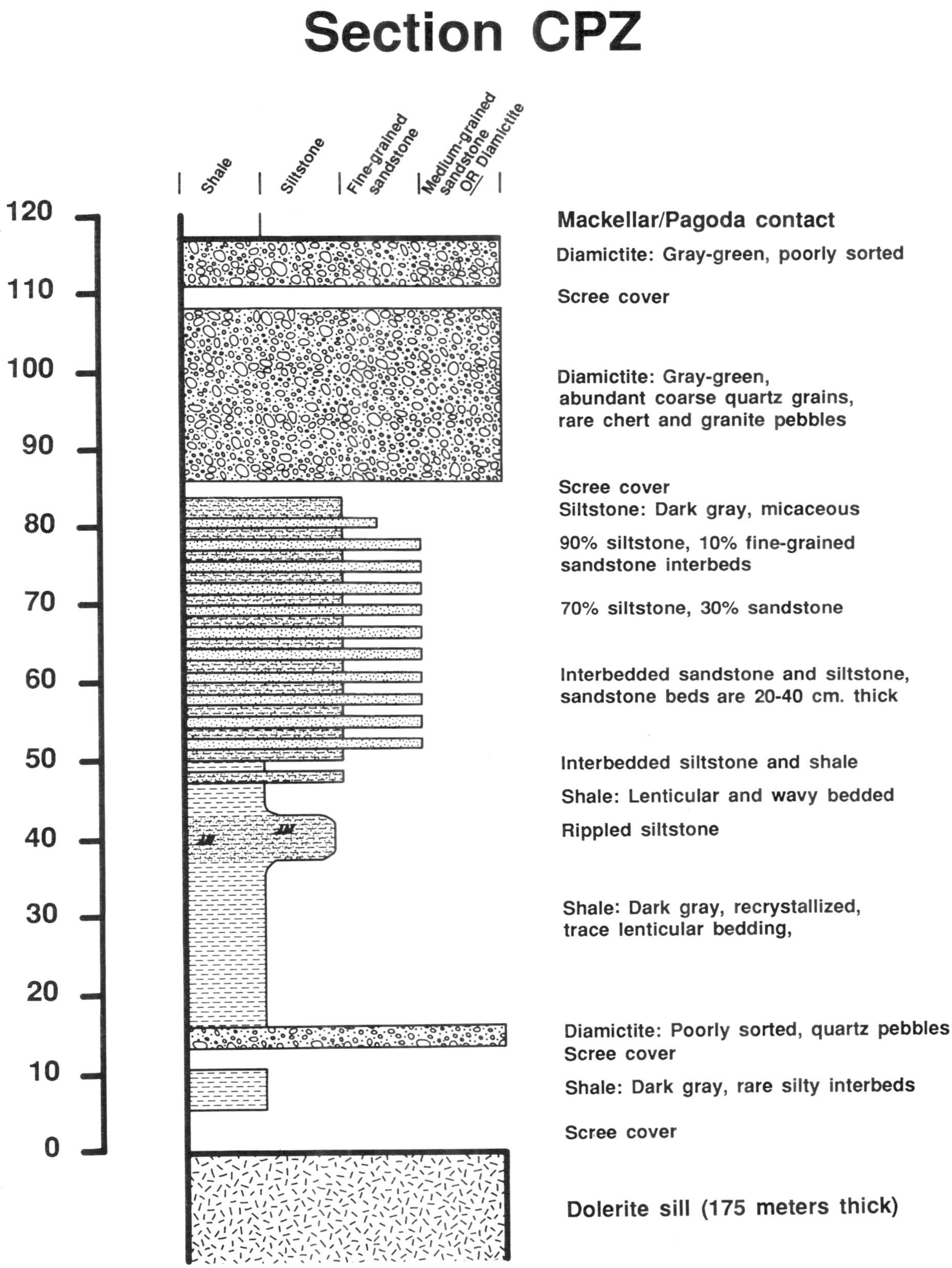

Fig. 5. Measured section CPZ from the Pagoda Formation, Clarkson Peak, illustrating the fine-grained end-member in the range of glacial deposits. Fine-grained sediments indicate deposition in proglacial ponds. (See Figure 3 for the key to the symbols.)

supply of meltwater. Preservation of standing water deposits was favored during the glacial retreats [*Miller*, 1989] because continued basin subsidence moved sediments below base level prior to the next advance. The thick shale at section CPZ suggests that the ice sheet was unstable and that retreat episodes occasionally persisted for prolonged periods during Pagoda deposition. Coarser-grained diamictite deposits that represent glacial advance cycles are not observed in the lower portion of this particular section.

Pagoda depositional environments can be summarized as glacial, although the variety of lithologies present necessitates more detailed interpretations. Diamictites deposited by moving ice are predominant in the Pagoda Formation, whereas the less abundant sandstones, siltstones, and shales are important for detailed interpretations and may be volumetrically important in specific sections.

Mackellar Formation

The Mackellar Formation is composed of fine- to medium-grained clastics that are interpreted to record the rapid postglacial infilling of an isostatically depressed basin [*Barrett et al.*, 1986; *Frisch*, 1987; *Miller et al.*, 1987]. Observations made during our fieldwork support this interpretation and are summarized here. Dominant lithologies in the Mackellar are sandstone, siltstone, and shale. The relative proportions of each lithology in the 13 measured sections are shown in Table 2.

Sandstones in the Mackellar Formation are present as thick sandstone beds, lenticular beds, and small-scale discontinuous features enclosed within finer-grained units. Thick sandstone beds are greater than 2 m thick and contain less than 10% silty interbeds.

TABLE 2. Lithologic Summary of the Mackellar Formation, Beardmore Glacier Region, Antarctica

Section	Lithology		
	Shale	Siltstone	Sandstone
MKR	4.7 m	21.4 m	46.7 m
MMQ	6.0 m	26.7 m	27.3 m
MMA	11.5 m	39.4 m	22.4 m
MMC	27.8 m	26.0 m	42.9 m
MMD	26.8 m	12.8 m	66.4 m
TRM	11.5 m	12.8 m	36.7 m
CPZ	22.3 m	26.2 m	22.1 m
MTM	8.3 m	41.7 m	30.4 m
TGA	21.9 m	18.1 m	37.0 m
MBO	22.6 m	32.6 m	44.8 m
CHI	10.2 m	70.9 m	75.3 m
MDK	31.4 m	21.2 m	66.4 m
Total	205.0 m	349.8 m	518.4 m
Percentage	19%	33%	48%

Sedimentary structures in the thick sandstones include trough cross-bedding up to 30 cm thick, ripple cross lamination, and planar bedding up to 20 cm thick. Soft sediment deformation is common, especially in the form of slumped or contorted beds. Thick sandstone beds increase in abundance toward the top of most Mackellar sections, giving the Mackellar a coarsening-upward pattern when viewed at the scale of the entire formation. Smaller coarsening-upward and fining-upward cycles, each 2 to 5 m thick, are common within this large-scale coarsening-upward pattern.

Thick sandstone bodies are interpreted in this study to have been delta lobes. Coarser-grained delta lobes become more common toward the top of the formation (Figure 6), indicating that small deltas prograded across the shallow Mackellar basin as the basin infilled with sediment. The rare lobes of coarser clastics which are found in the middle or lower portions of some Mackellar sections (Figures 7 and 8) have two possible interpretations. Localities that were near the margin of the Mackellar basin may have been influenced by early episodes of gradual lobe migration and deltaic deposition, whereas more distal locations did not receive coarse-grained delta clastics until the delta systems migrated toward the basin center later in Mackellar time. Section CPZ (Figure 7) is interpreted as a proximal deltaic deposit that has coarsening-upward sequences in the lower, middle, and upper portions of the measured section. A second interpretation for coarser clastics in the lower portions of the Mackellar Formation is delta bypassing. Disconnected thick sand bodies that do not appear to be part of a coarsening-upward sequence (Figure 8) may have been deposited when coarse sediments bypassed the delta top. Enhanced offshore transport caused by storms or sudden channel avulsion may occasionally have promoted the deposition of coarse clastics on the lower delta front or prodelta.

Other sandstone bodies are lenticular, have erosional bases, and contain trough cross-bedding, asymmetric ripple marks, rare planar tabular cross-bedding, and current lineation. These beds range from 50 cm to 5 m in thickness, and they are laterally continuous for distances of tens to hundreds of meters. Lenticular sandstone beds are interpreted as subaqueous distributary channels that fed the Mackellar delta lobes [*Miller et al.*, 1987].

Sandstones are also found as thin interbeds 1 to 10 cm thick, encased in the predominantly fine-grained portions of the Mackellar Formation. These thin sandstones occur as starved ripples, lenticular pods, and thin, discontinuous, wavy beds. Thin sandstone interbeds are interpreted to have been deposited on the delta front and record increased sedimentation rates during pulses of stronger current activity. Large-scale graded bedding and slump structures are rare or absent in these intervals, suggesting that the Mackellar delta front had a very low slope and relatively high stability [*Miller et al.*, 1987]. The thin Mackellar sandstones are interpreted to have been deposited from episodic,

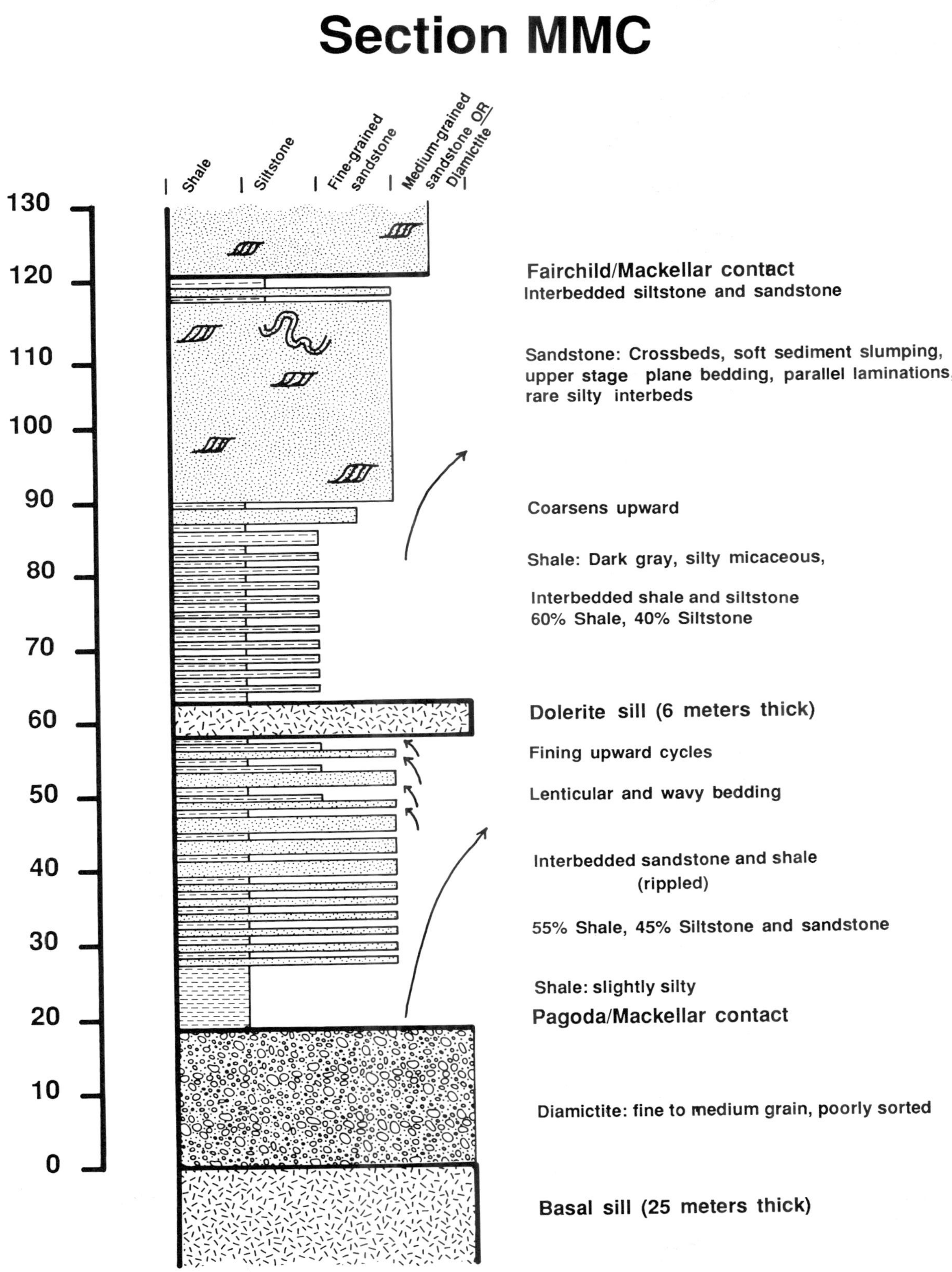

Fig. 6. Measured section MMC from the Mackellar Formation, showing the large-scale coarsening-upward pattern that is typical of that formation. Smaller-scale coarsening-upward and fining-upward sequences are visible within the section. (See Figure 3 for the key to the symbols.)

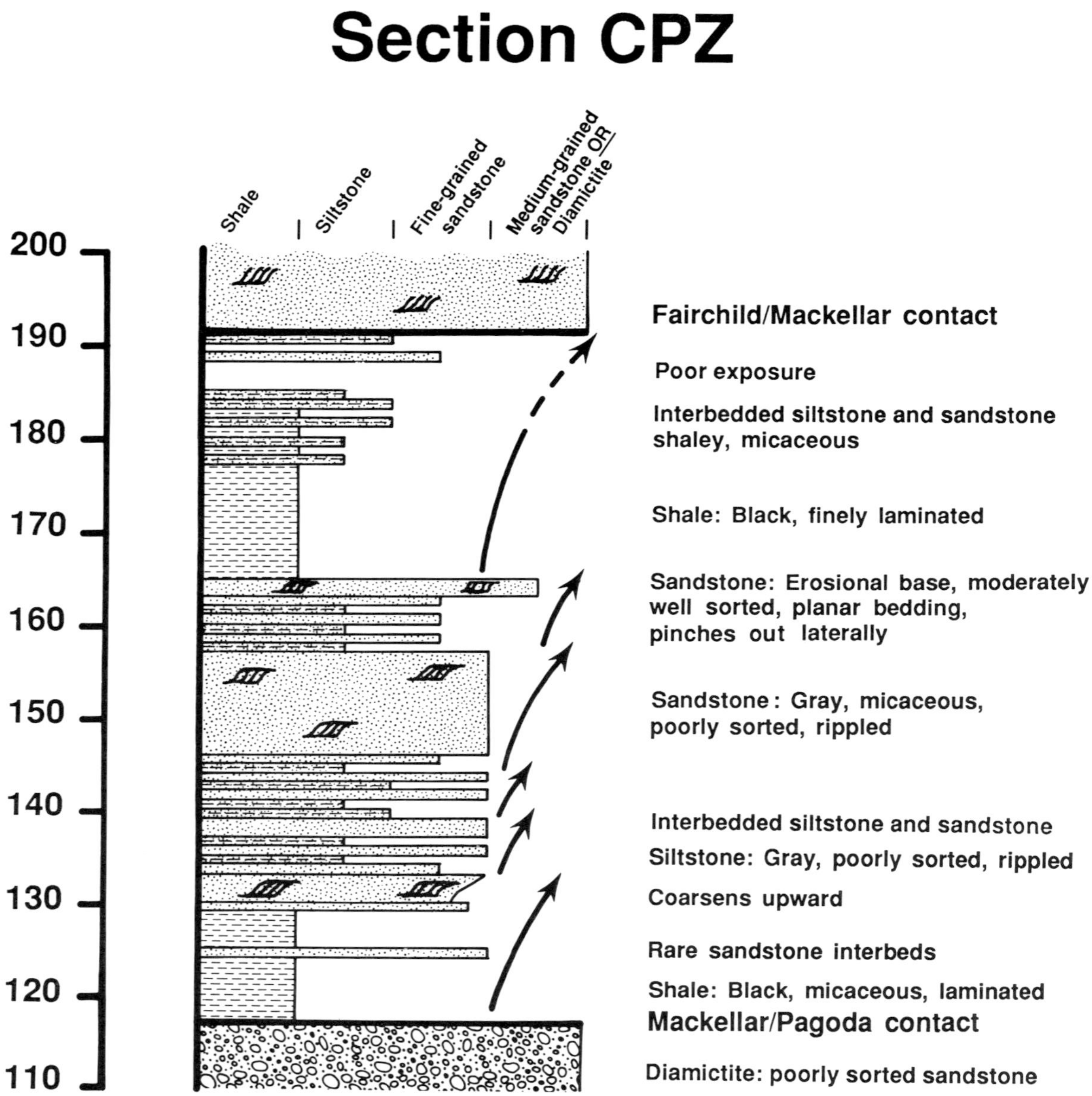

Fig. 7. Measured section CPZ from the Mackellar Formation. The arrows indicate small-scale coarsening-upward sequences. The thick sand body at 147–158 m is interpreted as an abandoned delta lobe. (See Figure 3 for the key to the symbols.)

distal, low-density mass flows in a broad, shallow, subaqueous basin.

Siltstones in the Mackellar Formation also indicate rapid deposition and an abundant sediment supply. Sedimentary structures include starved ripples, wavy and flaser bedding (Figure 9), linguoid ripples, climbing ripples, rare oscillation ripples, and rare interference ripples. The siltstones are generally interbedded with shales and/or sandstones, indicating rapid changes in current strength and sediment supply. Lateral continuity of the siltstones is highly variable, and bed thicknesses range from millimeters to tens of centimeters. The combination of high sediment supply and fluctuations in current velocity resulted in rapid deposition of the silty clastics. We interpret these siltstones to be part of the proximal delta-front complex.

Shales in the Mackellar Formation occur either as thin interbeds in the upper part of the formation or as thick beds at the base of the formation. The thin shales in the upper Mackellar commonly exhibit fine- to medium-scale laminae, are silty, and may contain

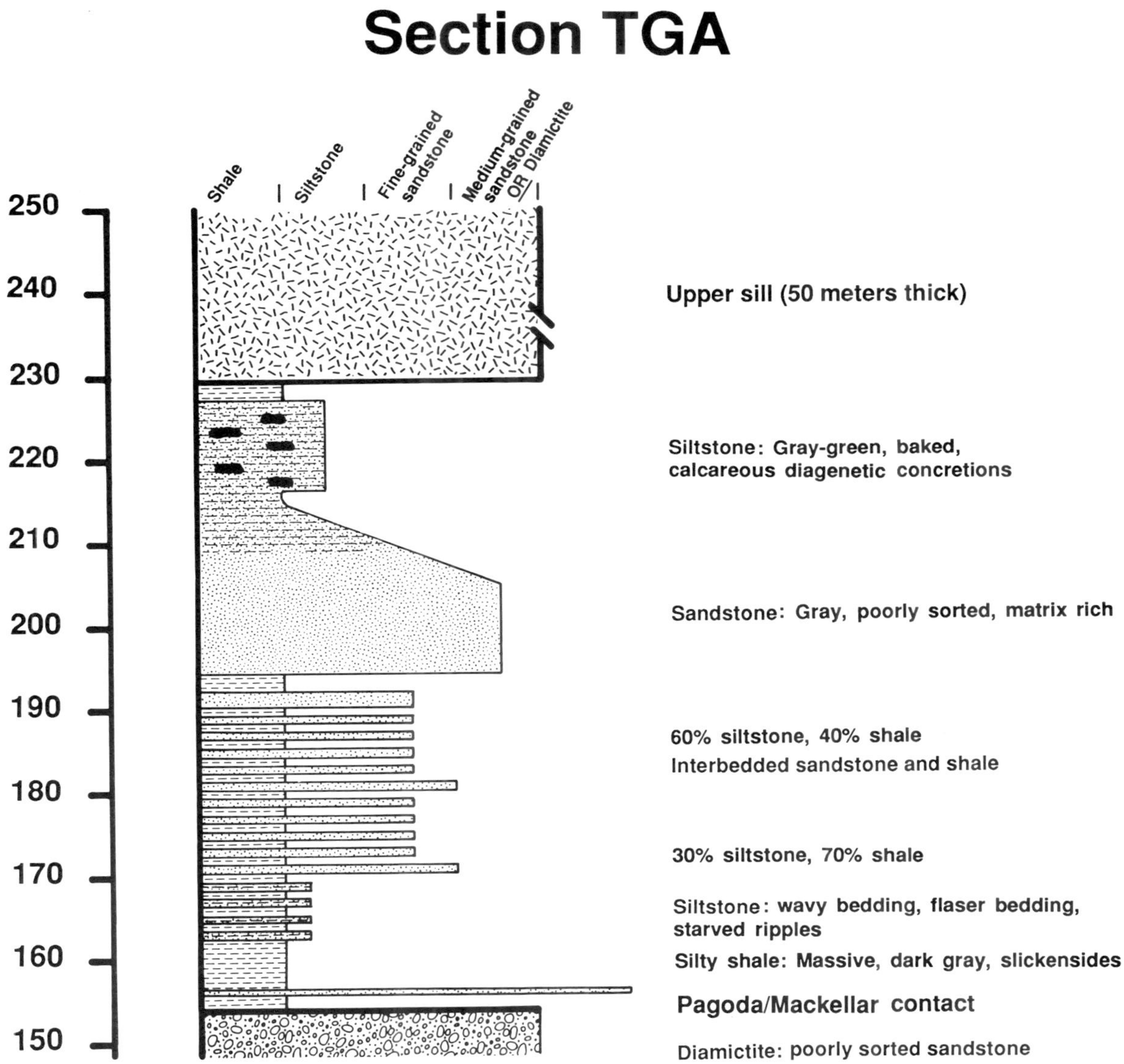

Fig. 8. Measured section TGA from the Mackellar Formation, showing the large-scale coarsening-upward sequence, followed by delta lobe abandonment and a return to quiet or deepwater conditions. (See Figure 3 for the key to the symbols.)

small-scale ripple cross lamination. These silty shales are interpreted to be delta-front deposits within the Mackellar basin and represent quiet water deposition with sporadic episodes of more rapid sedimentation.

Thick basal shales in the Mackellar Formation are dark gray to black, are laminated, and occasionally exhibit silty laminations and rare silty interbeds. The laminations are interpreted to record offshore movement by underflows which reworked and winnowed the sediments and transported oxygen into the deeper basin. These underflows are important to the interpretation of bottom-water conditions in the Mackellar basin because deltaic underflows may supply oxygen to basinal sediments [*Ettensohn and Elam*, 1985]. The continued influx of oxygen- and sediment-laden currents from the Mackellar delta system prevented the development of anaerobic conditions even though thick basal shales accumulated during the initial phases of Mackellar deposition. Basal shale intervals measured in this study range from 2 to 27 m in thickness and exhibit a consistent regional thickness pattern that may define the original shape of this portion of the Mackellar depositional basin (Figure 10). Thick shales are interpreted as starved basin deposits, located farther from shore in depressions where quiet water conditions persisted the longest before the onset of deltaic sedimentation.

Fig. 9. Wavy and flaser bedding in delta-front siltstones and sandstones from the Mackellar Formation, section TGA. The Swiss army knife is shown for scale.

Isopach patterns of total Mackellar thickness also help to define the shape of the basin in this study area (Figure 11), and paleocurrent patterns [*Isbell,* 1990] support the hypothesis that sediment was supplied from several sides of an elongate depression. The shape outlined by these data is not necessarily the shape of the entire Mackellar basin but may represent an isolated arm or inlet along the interior waterway that filled the foreland basin described by *Elliot* [1975] and *Collinson* [1990].

The lithologic associations and sedimentary structures described here support the interpretation that the Mackellar Formation was deposited while sediment-laden streams entered a standing body of water. A combination of isostatic rebound and high sedimentation rates allowed marginal deltas to prograde over and finally infill the shallow Mackellar basin.

Fairchild Formation

The Fairchild Formation is dominated by sheet-form bodies of medium- to coarse-grained sandstone that were deposited on a broad, flat braidplain (Table 3) [*Barrett,* 1968; *Barrett et al.,* 1986; *Isbell,* 1990]. Fine-grained sediments are rare in the Fairchild Formation, so it was not examined to the same extent as the other formations in this study. Sedimentary structures observed in the sandstone beds include large-scale cross-bedding with bed sets up to 1 m thick, ripple cross laminae, climbing ripples, upper flow regime parallel lamination, and soft sediment slump structures up to 5 m thick. Sandstone sheets and individual sandstone beds have erosional bases that are concave upward and erosional boundaries that intersect each other laterally. Quartz pebbles are common at some localities as a lag deposit at the base of the Fairchild Formation. The sandstone sheets and the individual bed sets within these sheets are interpreted as braided stream deposits [*Barrett,* 1968;

TABLE 3. Lithologic Summary of the Fairchild Formation, Beardmore Glacier Region, Antarctica

Section	Lithology		
	Shale	Siltstone	Sandstone
MKR	...	0.5 m	17.2 m
MMD	1.0 m	...	162.0 m
TRM	1.0 m	3.9 m	53.7 m
CPZ	...	0.6 m	19.4 m
Total	2.0 m	5.0 m	252.3 m
Percentage	<1%	2%	97%

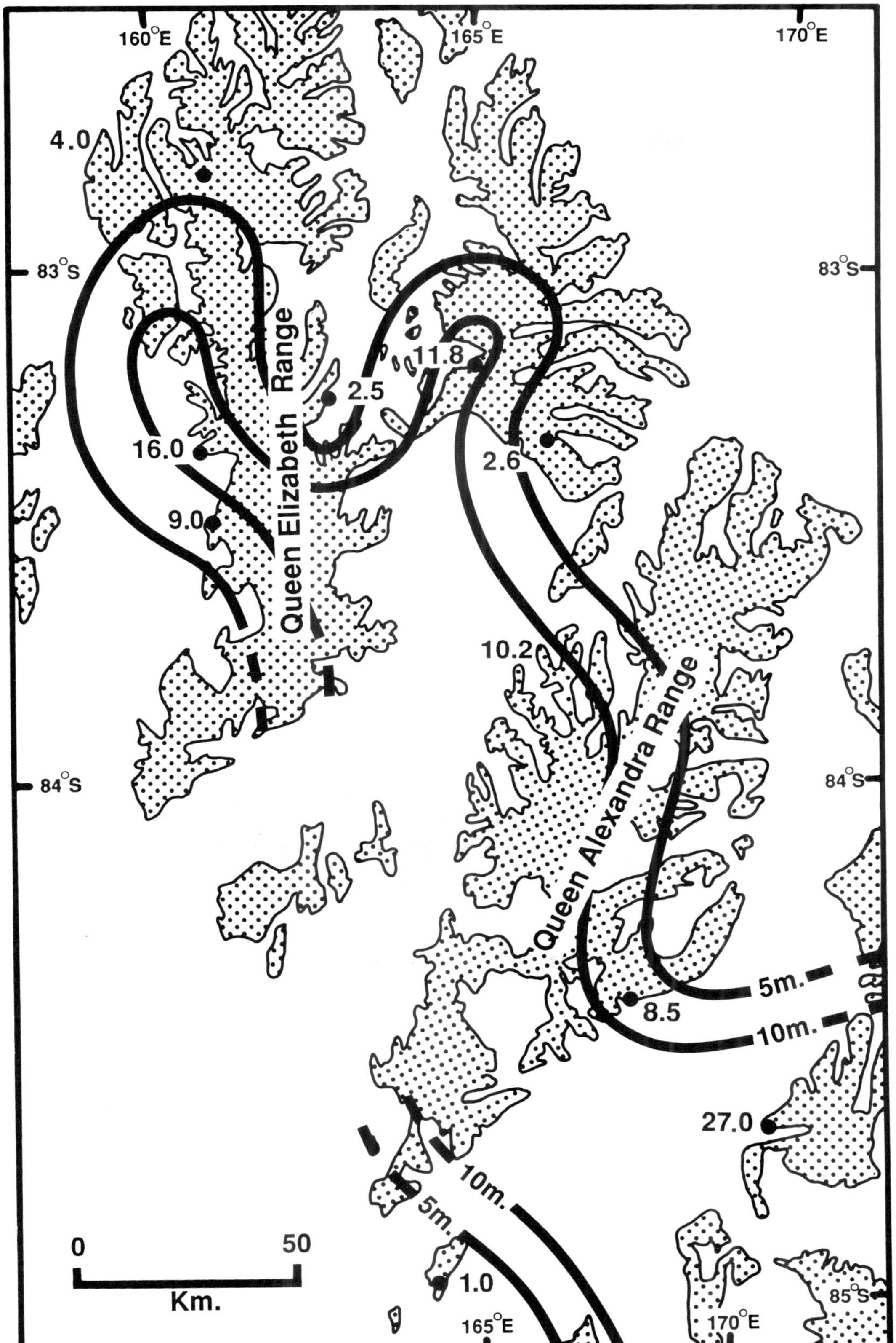

Fig. 10. Contour map showing the thickness of the basal shale in the Mackellar Formation. Thicker shale deposits are interpreted to indicate deeper water and increasing distance from shore.

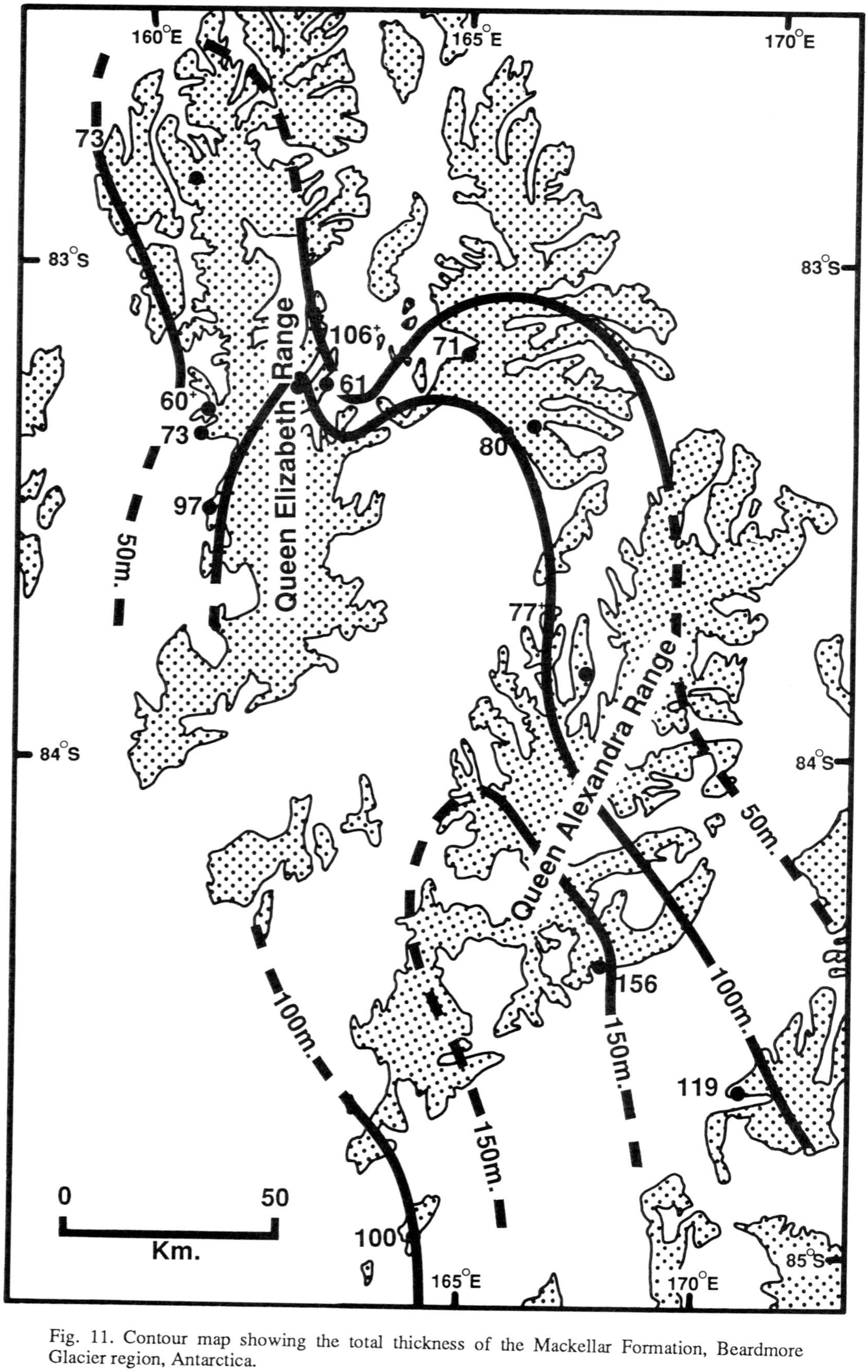

Fig. 11. Contour map showing the total thickness of the Mackellar Formation, Beardmore Glacier region, Antarctica.

Isbell, 1990]. Multiple stacked channel sequences that formed as low-gradient braided streams migrated across the broad, flat plain that formed when the Mackellar deltaic complex infilled the postglacial basin.

Shales and siltstones form less than 3% of the Fairchild Formation, and their sedimentary structures include thin to thick laminae, current ripples, and wavy bedding. These fine-grained clastics are interpreted as overbank deposits, which were rarely preserved in the Fairchild Formation [*Isbell*, 1990]. Continual reworking destroyed the finer-grained facies and preferentially preserved the coarser channel sands. For a complete description of the Fairchild Formation see *Isbell* [1990].

Buckley Formation

The Buckley Formation is composed of sandstones, siltstones, shales, and coals that were deposited in braided stream systems, located on what may have been the landward margin of a developing foreland basin [*Isbell*, 1990; *Collinson*, 1990]. Continued subsidence of the basin preserved a complex group of sediments, including finer-grained lithologies that represent a variety of depositional environments (Table 4). Sedimentary structures and lithologic associations can be used to identify several depositional environments within the Buckley Formation.

TABLE 4. Lithologic Summary of the Buckley Formation, Beardmore Glacier Region, Antarctica

Section	Lithology			
	Shale	Siltstone	Sandstone	Coal
MMD	42.2 m	10.2 m	181.8 m	10.0 m
MPI	77.6 m	119.1 m	148.2 m	21.9 m
MTR	8.3 m	119.1 m	160.4 m	18.3 m
MAR	113.4 m	97.9 m	136.4 m	19.6 m
LPP	4.5 m	24.8 m	36.2 m	2.5 m
MBO	3.0 m	25.9 m	102.2 m	7.9 m
Total	249.2 m	367.2 m	759.8 m	80.2 m
Percentage	17%	25%	52%	6%

The lower portion of the Buckley Formation is dominated by thick sandstone beds [*Barrett*, 1968; *Barrett et al.*, 1986; *Isbell*, 1990]. These beds are less common in the upper portion of the Buckley Formation, although they remain an important component. Sandstone beds range from 50 cm to 15 m in thickness and occur both as sheets and as channel-form bodies. Channel-form bodies have erosional, concave upward bases with lateral extent of tens to hundreds of meters. Sheet sands have undulating lower surfaces and commonly extend for thousands of meters. Sedimentary structures in the thick sandstone beds include trough cross-beds up to 1 m high, tabular planar cross-bed sets 10 to 30 cm thick, and asymmetrical ripple marks. Large- and small-scale fining-upward sequences are common. Small-scale fining-upward sequences are centimeters to decimeters thick and occur as either a subtle upward decrease in grain size or a systematic upward decrease in the size of the sedimentary structures. Small-scale fining-upward sequences are interpreted to be evidence for decreasing flow velocity due to decreasing discharge or migration of the channel axis. Large-scale fining-upward sequences are up to tens of meters thick, with thick sandstone at the base and a progression to finer-grained lithologies upward. These sandstone bodies are interpreted as braided stream channels and multilayered channel complexes and may include associated crevasse splays [*Isbell*, 1990].

Siltstones, shales, and coals exhibit lithologic associations that are important for interpreting depositional environments within the Buckley Formation. The large-scale fining-upward sequences discussed above generally grade into siltstones, shales, and coals. These siltstones and shales contain ripple marks, mud cracks, plant fossils, and paleosol horizons (Figure 12) which indicate the presence and preservation of floodplains near the coarser channel systems. Other fine-grained units do not fine upward but instead contain laminated and interbedded layers of siltstone and shale. Laminated units are interpreted as ponded deposits and represent low-lying areas near the Buckley river systems [*Isbell*, 1990].

Coal beds in the Buckley Formation are both allochthonous and autochthonous [*Isbell*, 1990]. Allochthonous coal beds range from 5 cm to more than 1 m in thickness and are encased in shale, siltstone, or sandstone beds. Allochthonous coals pinch out laterally within 2 to 200 m. These thin coal beds often contain silty or shaley interbeds and formed in low-lying areas where organic matter accumulated with an occasional clastic influx [*Isbell*, 1990].

In contrast to the allochthonous coal beds, autochthonous coal beds range from 10 cm to more than 4 m in thickness and are commonly underlain by rooted horizons. Autochthonous coal beds cap fining-upward sequences (Figure 13) and are laterally traceable for hundreds to thousands of meters along the well-exposed outcrops in the Beardmore Glacier area. Autochthonous coals formed in broad lowlands or along the margins of lakes that were distal to the river channel and were therefore infrequently influenced by floods and clastic influx [*Isbell*, 1990].

In summary, the Buckley Formation was deposited in a variety of environments within a braided stream setting. Braided stream channel deposits predominate in the lower Buckley, but deposits of finer-grained floodplain and peat-forming environments become more abundant in the upper Buckley. Sedimentary structures and facies associations allow the distinction of these different environments.

Fig. 12. Rooted horizon within the Buckley Formation, from Lamping Peak, section LPP, illustrating paleosol development. Arrows point to thin root traces within the siltstone bed. The Swiss army knife is shown for scale.

ORGANIC CARBON

The factors that control the occurrence of organic carbon in sedimentary rocks must be examined before organic carbon contents can be properly interpreted. The first part of this section is a discussion of the relative importance of the different factors that control the input and preservation of organic carbon in the four studied formations. After the major controlling factors have been identified, organic carbon contents from Carboniferous/Permian sediments in the Victoria Group are used for two different purposes. First, a combination of vitrinite reflectance, field relations, and TOC data are used to demonstrate the utility of carbon compounds as indicators of alteration in thermally modified strata. Second, once the extent of thermal alteration is known, the characteristics of relatively unaltered samples are combined with field observations to develop general interpretations about depositional environments and organic productivity in the Beardmore Glacier region during the late Paleozoic.

Factors Influencing the Amounts of Organic Carbon in Beacon Sediments

Several factors control the occurrence of organic matter in sedimentary rocks. These factors include grain size, type of organic matter, level of primary productivity, amount of bioturbation, dissolved oxygen content, extent of bacterial degradation, carbon contribution from recycling, sedimentation rates, and extent of postdepositional alteration [*Demaison and Moore,* 1980; *Jones,* 1983]. Organic carbon content in the resulting sedimentary rock is a balance between factors that control the input of organic matter and factors that control the destruction of organic matter. The importance of these factors in Beacon sediments must be evaluated before the results of organic carbon analyses can be interpreted.

Organic carbon content has generally been observed to increase as sediment grain size decreases [*Jones,* 1983]. Within the present study, this effect has been minimized by sampling only the finest-grained intervals from each formation. Samples that are used

Fig. 13. Coal bed in the Buckley Formation, section MAR, capping a fining-upward sequence. A bamboo pole (1.5 m long) was placed at that location for scale.

for organic carbon analyses are either shales or siltstones (Table 5). This method of limiting grain size effects is imperfect, but the sediments of the Beacon Supergroup are too highly indurated for conventional methods of disaggregation and mechanical size separation to be effective. Within these limits, every effort has been made to ensure that samples from each of the four formations are consistently fine grained.

Because different types of organic matter have different susceptibilities to degradation, the preservation potential of the TOC is dependent on the type of organic matter present. Kerogen types were identified by a commercial laboratory using reflected light microscopy. Samples from nine measured sections were analyzed, including two samples from the Pagoda Formation, 24 samples from the Mackellar Formation, and 11 samples from the Buckley Formation. The averaged analyses show that the kerogens (insoluble organic matter) present are relatively uniform, with vitrinite and inertinite as the dominant kerogen types in each formation (Figure 14). Vitrinite and inertinite are the insoluble organic carbon compounds that form when terrestrial organic matter undergoes burial and decomposition [*Tissot and Welte,* 1984]. As a result, the major source of carbon to these sediments is consistently terrestrial, which eliminates the possibility that changes in the types of organic carbon present are responsible for the differences in organic carbon contents.

TABLE 5. Results From Organic Carbon Analyses, Lithologic Determination, and Vitrinite Reflectance Analysis

Sample ID*	Formation	Total Carbon, %	Inorganic Carbon, %	Organic Carbon, %	Lithology	Vitrinite Reflectance	Comment
MKR 6.0	Mackellar	0.44	0.00	0.44	siltstone	0.78	oxidized
MKR 12.4	Mackellar	0.30	0.10	0.20	silty shale		
MKR 23.2	Mackellar	0.33	0.02	0.31	shaley siltstone		
MKR 34.3	Mackellar	0.35	0.02	0.33	silty shale		
MKR 42.2	Mackellar	0.37	0.03	0.34	silty shale	0.96	
MKR 58.5	Mackellar	0.35	0.01	0.34	siltstone	1.26	
MKR 90.5	Fairchild	0.02	0.00	0.02	siltstone		
MMQ 4.3	Mackellar	0.58	0.01	0.57	silty shale	2.38	
MMQ 26.5	Mackellar	0.26	0.00	0.26	sandy siltstone		
MMQ 29.0	Mackellar	0.16	0.01	0.15	shale		
MMQ 33.0	Mackellar	0.19	0.00	0.19	shale	2.34	
MMA 6.6	Mackellar	0.04	0.01	0.03	silty shale	...	barren
MMA 14.2	Mackellar	0.01	0.00	0.01	sandy siltstone		
MMA 24.3	Mackellar	0.02	0.00	0.02	sandy siltstone		
MMA 38.8	Mackellar	0.07	0.00	0.07	sandy siltstone		
MMA 51.0	Mackellar	0.10	0.01	0.09	siltstone		
MMA 59.8	Mackellar	0.10	0.00	0.10	siltstone	3.02	coked
MMA 73.2	Mackellar	0.35	0.00	0.35	sandy siltstone	3.44	
MMA 87.0	Fairchild	0.87	0.00	0.87	silty shale	3.21	
MMP 1.5	Pagoda	0.11	0.01	0.10	sandy siltstone	...	coked
MMP 23.5	Pagoda	0.11	0.00	0.11	sandy siltstone		
MMP 74.5	Pagoda	0.04	0.01	0.03	shale	...	barren
MMC 17.6	Pagoda	0.07	0.00	0.07	sandy siltstone		
MMC 18.7	Mackellar	0.03	0.00	0.03	shale		
MMC 20.5	Mackellar	0.05	0.00	0.05	silty shale		
MMC 29.0	Mackellar	0.08	0.00	0.08	shale		
MMC 37.7	Mackellar	0.07	0.01	0.06	silty shale		
MMC 49.1	Mackellar	0.12	0.00	0.12	sandy siltstone	2.84	
MMC 56.7	Mackellar	0.06	0.01	0.06	siltstone		
MMC 63.7	Mackellar	0.07	0.00	0.07	shale		
MMC 64.4	Mackellar	0.19	0.00	0.19	shale		
MMC 71.0	Mackellar	0.14	0.01	0.13	siltstone		
MMC 74.8	Mackellar	0.15	0.01	0.14	shale	...	barren
MMC 82.8	Mackellar	0.18	0.00	0.18	shale		
MMC 87.8	Mackellar	0.22	0.00	0.22	shaley siltstone		
MMC 95.2	Mackellar	0.25	0.00	0.25	silty shale		
MMC 117.5	Mackellar	0.29	0.02	0.27	shale	3.00	
MMD 7.5	Mackellar	0.53	0.00	0.53	sandy siltstone		
MMD 11.0	Mackellar	0.38	0.00	0.38	shale	2.02	
MMD 27.0	Mackellar	0.94	0.30	0.64	silty shale	1.93	
MMD 62.5	Mackellar	8.32	8.14	0.18	siltstone		
MMD 75.5	Mackellar	0.39	0.01	0.38	silty shale	2.29	
MMD 106.0	Mackellar	0.47	0.00	0.47	silty shale		
MMD 115.5	Fairchild	0.43	0.00	0.43	silty shale	2.11	
MMD 270.0	Buckley	4.94	0.00	4.94	sandy siltstone	0.70	
MMD 283.5	Buckley	5.31	0.00	5.31	sandy siltstone	0.93	
MMD 295.5	Buckley	26.60	0.00	26.60	silty shale	0.38	
MMD 318.5	Buckley	40.52	0.00	40.52	coal	0.50	
MMD 384.0	Buckley	58.05	0.00	58.05	coal	0.48	
MMD 417.0	Buckley	12.70	0.00	12.70	silty shale	1.01	
MMD 440.5	Buckley	13.33	0.00	13.33	shaley coal	2.09	
TRM 0.0	Mackellar	0.50	0.06	0.44	silty shale	1.21	
TRM 11.5	Mackellar	0.53	0.09	0.44	shaley siltstone		
TRM 16.5	Mackellar	0.65	0.03	0.62	silty shale		
TRM 27.5	Mackellar	0.43	0.02	0.41	silty shale	1.37	
TRM 35.1	Mackellar	0.45	0.02	0.43	shaley siltstone		
TRM 47.1	Mackellar	0.36	0.05	0.31	siltstone		
TRM 71.6	Fairchild	0.56	0.01	0.55	silty sandstone	1.66	

TABLE 5. (continued)

Sample ID*	Formation	Total Carbon, %	Inorganic Carbon, %	Organic Carbon, %	Lithology	Vitrinite Reflectance	Comment
CPZ 16.3	Pagoda	0.03	0.00	0.03	silty shale		
CPZ 46.1	Pagoda	0.09	0.00	0.09	silty shale		
CPZ 117.9	Mackellar	0.20	0.00	0.20	shale		
CPZ 123.4	Mackellar	0.19	0.00	0.19	silty shale	3.10	coked
CPZ 125.1	Mackellar	0.22	0.00	0.22	shale		
CPZ 137.0	Mackellar	0.20	0.00	0.20	siltstone		
CPZ 145.9	Mackellar	0.28	0.00	0.28	sandy siltstone	2.64	
CPZ 158.1	Mackellar	0.24	0.00	0.24	shale		
CPZ 166.5	Mackellar	0.22	0.00	0.22	shale		
CPZ 176.1	Mackellar	0.19	0.04	0.15	silty shale	2.78	
CPZ 183.4	Mackellar	0.21	0.02	0.19	siltstone		
CPZ 190.5	Mackellar	0.20	0.00	0.20	silty shale		
CPZ 201.8	Fairchild	5.74	5.59	0.15	siltstone		
MTM 129.0	Pagoda	0.24	0.05	0.19	silty shale	2.81	coked
MTM 167.0	Pagoda	0.32	0.00	0.32	shaley siltstone		
MTM 186.2	Pagoda	0.53	0.15	0.38	shale	3.18	
MTM 213.5	Pagoda	0.36	0.02	0.34	shale	...	barren
MTM 216.7	Mackellar	0.58	0.30	0.28	shaley siltstone		
MTM 223.6	Mackellar	0.47	0.10	0.36	siltstone		
MTM 232.3	Mackellar	0.43	0.14	0.29	sandy siltstone		
MTM 246.2	Mackellar	0.46	0.00	0.46	siltstone	2.82	coked
MTM 269.0	Mackellar	0.42	0.00	0.42	silty shale	2.99	
MTM 279.1	Mackellar	0.34	0.04	0.30	siltstone		
MTM 294.2	Mackellar	0.30	0.02	0.28	siltstone		
TGA 1.0	Pagoda	0.05	0.00	0.05	silty shale		
TGA 31.0	Pagoda	0.04	0.00	0.04	sandy siltstone		
TGA 41.0	Pagoda	0.19	0.14	0.05	sandy siltstone	3.40	
TGA 86.0	Pagoda	0.16	0.00	0.16	silty shale		
TGA 96.7	Pagoda	0.26	0.00	0.26	silty shale	3.07	
TGA 151.7	Mackellar	0.33	0.00	0.33	shale	3.10	
TGA 154.9	Mackellar	0.26	0.00	0.26	silty shale		
TGA 156.5	Mackellar	0.27	0.01	0.26	silty shale		
TGA 160.0	Mackellar	0.25	0.01	0.24	shale	...	coked
TGA 162.0	Mackellar	0.24	0.00	0.24	shale		
TGA 163.5	Mackellar	0.20	0.00	0.20	sandy siltstone		
TGA 163.6	Mackellar	0.14	0.01	0.13	sandy siltstone		
TGA 170.0	Mackellar	0.14	0.02	0.12	sandy siltstone		
TGA 173.0	Mackellar	0.17	0.00	0.17	sandy siltstone	3.70	coked
TGA 194.0	Mackellar	0.11	0.00	0.11	sandy siltstone		
TGA 200.3	Mackellar	0.07	0.00	0.07	siltstone		
TGA 211.0	Mackellar	0.10	0.01	0.09	siltstone		
TGA 213.5	Mackellar	0.09	0.00	0.09	silty shale		
TGA 227.5	Mackellar	0.04	0.00	0.04	silty shale		
TGA 228.0	Mackellar	0.07	0.00	0.07	shale		
TGF 0.0	Mackellar	0.04	0.01	0.03	shale	...	barren
TGF 10.0	Mackellar	0.04	0.00	0.04	silty shale		
TGF 136.0	Fairchild	0.43	0.00	0.43	siltstone	...	coked
TGF 138.5	Fairchild	0.55	0.00	0.55	shale		
TGF 218.0	Buckley	0.20	0.01	0.19	shale	...	coked
WLG 5.0	Buckley	1.35	0.18	1.17	coaly shale	2.34	
WLG 30.0	Buckley	1.08	0.00	1.08	silty shale		
WLG 53.0	Buckley	1.07	0.00	1.07	silty shale		
WLG 58.0	Buckley	27.11	0.00	27.11	shale	1.91	
MPI1 28.6	Buckley	0.95	0.00	0.95	shale	6.23	coked
MPI1 35.8	Buckley	6.46	0.00	6.46	shaley coal		
MPI1 53.1	Buckley	1.58	0.00	1.58	shale	2.50	coked
MPI1 54.0	Buckley	6.84	0.01	6.83	shaley coal	3.26	coked
MPIu 40.4	Buckley	0.78	0.00	0.78	shale	2.53	

TABLE 5. (continued)

Sample ID*	Formation	Total Carbon, %	Inorganic Carbon, %	Organic Carbon, %	Lithology	Vitrinite Reflectance	Comment
MPIu 52.8	Buckley	2.26	0.00	2.26	silty shale		
MPIu 117.3	Buckley	3.32	0.00	3.32	shale	3.15	
MPIu 159.4	Buckley	29.97	0.00	29.97	coaly shale	2.57	
MPIu 183.8	Buckley	37.73	0.00	37.73	coaly shale	2.65	
MPIu 214.3	Buckley	2.83	0.00	2.83	shale	3.87	
MPIu 222.9	Buckley	9.21	0.00	9.21	shale		
MPIu 241.5	Buckley	1.23	0.00	1.23	shale	3.25	
MTR 13.1	Buckley	3.57	0.01	3.56	coaly shale	4.36	
MTR 61.1	Buckley	6.95	0.00	6.95	silty shale	3.31	
MTR 121.7	Buckley	29.49	0.00	20.49	coal		
MTR 130.5	Buckley	5.24	0.00	5.24	shale	3.13	
MTR 159.5	Buckley	8.00	0.00	8.00	coaly shale	2.41	
MTR 185.9	Buckley	9.52	0.00	9.52	shaley coal		
MTR 203.7	Buckley	22.00	0.00	22.00	coaly shale	1.39	
MAR1 14.4	Buckley	0.13	0.00	0.13	silty shale	...	coked
MAR1 45.2	Buckley	1.83	0.01	1.82	coaly shale	2.42	
MAR1 108.2	Buckley	29.61	0.00	29.61	coaly shale		
MAR1 131.2	Buckley	14.92	0.00	14.92	shaley coal	3.46	
MARm 33.2	Buckley	0.06	0.00	0.06	silty shale		
MARm 51.7	Buckley	0.19	0.00	0.19	coaly shale	3.49	coked
MARm 95.8	Buckley	0.80	0.00	0.80	silty shale		
MARm 109.6	Buckley	3.66	0.01	3.65	coaly shale		
MARm 114.1	Buckley	0.96	0.00	0.96	shale	3.47	
MARm 152.0	Buckley	1.00	0.00	1.00	shale	4.94	
MARu 9.2	Buckley	1.80	0.00	1.80	silty shale	2.30	
MARu 16.0	Buckley	18.62	0.00	18.62	shaley coal		
MARu 37.0	Buckley	0.99	0.01	0.98	silty shale	2.93	
LPP 28.0	Buckley	0.64	0.01	0.63	shaley siltstone	2.96	
LPP 39.5	Buckley	0.40	0.00	0.40	silty shale		
LPP 54.5	Buckley	0.23	0.00	0.23	siltstone	2.67	
LPP 58.5	Buckley	51.39	0.00	51.39	coal		
MBO1 0.2	Pagoda	0.11	0.00	0.11	sandy siltstone		
MBO1 9.5	Pagoda	0.11	0.00	0.11	sandy siltstone		
MBO1 38.5	Mackellar	0.96	0.03	0.93	shaley siltstone	3.90	
MBO1 54.5	Mackellar	1.27	0.00	1.27	silty shale		
MBO1 67.8	Mackellar	1.27	0.00	1.27	siltstone	3.21	
MBO1 75.5	Mackellar	2.02	0.00	2.02	silty shale		
MBO1 88.0	Mackellar	2.34	0.00	2.34	shale	2.86	
MBO1 104.0	Mackellar	1.04	0.01	1.03	shaley siltstone	2.23	
MBO1 110.5	Mackellar	4.86	4.26	0.59	silty shale		
MBO1 125.0	Mackellar	1.09	0.00	1.09	shale	1.71	
MBO1 134.0	Mackellar	1.27	0.00	1.27	silty shale		
MBO1 170.0	Fairchild	2.81	0.01	2.80	shaley siltstone	2.23	
MBOu 27.5	Buckley	3.92	0.01	3.91	shale	1.50	
MBOu 89.8	Buckley	12.36	0.00	12.36	silty shale	1.40	
MBOu 116.1	Buckley	5.89	0.00	5.89	shale	1.32	
CHI 91.0	Pagoda	0.13	0.00	0.13	sandy siltstone		
CHI 108.8	Pagoda	0.32	0.00	0.32	sandy siltstone		
CHI 129.2	Mackellar	0.38	0.01	0.37	shaley siltstone		
CHI 132.6	Mackellar	0.38	0.01	0.37	silty shale	3.92	
CHI 139.4	Mackellar	0.29	0.00	0.29	shale		
CHI 153.0	Mackellar	0.17	0.00	0.17	sandy siltstone		
CHI 169.2	Mackellar	0.17	0.00	0.17	sandy siltstone		
CHI 178.5	Mackellar	0.43	0.00	0.43	shaley siltstone	2.24	coked
CHI 185.3	Mackellar	3.00	2.83	0.17	limey siltstone		
CHI 192.1	Mackellar	0.16	0.01	0.15	sandy siltstone		
MDK 0.5	Mackellar	0.17	0.00	0.17	silty shale		
MDK 10.7	Mackellar	0.31	0.00	0.31	silty shale	...	coked

TABLE 5. (continued)

Sample ID*	Formation	Total Carbon, %	Inorganic Carbon, %	Organic Carbon, %	Lithology	Vitrinite Reflectance	Comment
MDK 20.5	Mackellar	0.57	0.00	0.57	silty shale	5.44	
MDK 30.8	Mackellar	0.39	0.02	0.39	shaley siltstone		
MDK 40.9	Mackellar	0.11	0.00	0.11	sandy siltstone		
MDK 50.1	Mackellar	0.08	0.00	0.08	siltstone		
MDK 65.5	Mackellar	0.08	0.02	0.06	silty shale		
MDK 79.0	Mackellar	0.10	0.00	0.10	sandy siltstone		
MDK 91.5	Mackellar	0.08	0.01	0.07	sandy siltstone		
MDK 119.0	Mackellar	0.21	0.00	0.21	shaley siltstone	5.40	
MDK 209.0	Fairchild	0.11	0.00	0.11	sandy siltstone	4.81	coked
MDK 216.5	Fairchild	0.13	0.01	0.12	siltstone		
MDK 229.0	Fairchild	0.08	0.01	0.07	siltstone		

*This column lists the section designations and locations as shown in Figure 1. Numeric designations in the sample ID labels indicate the heights above the bases of the sections.

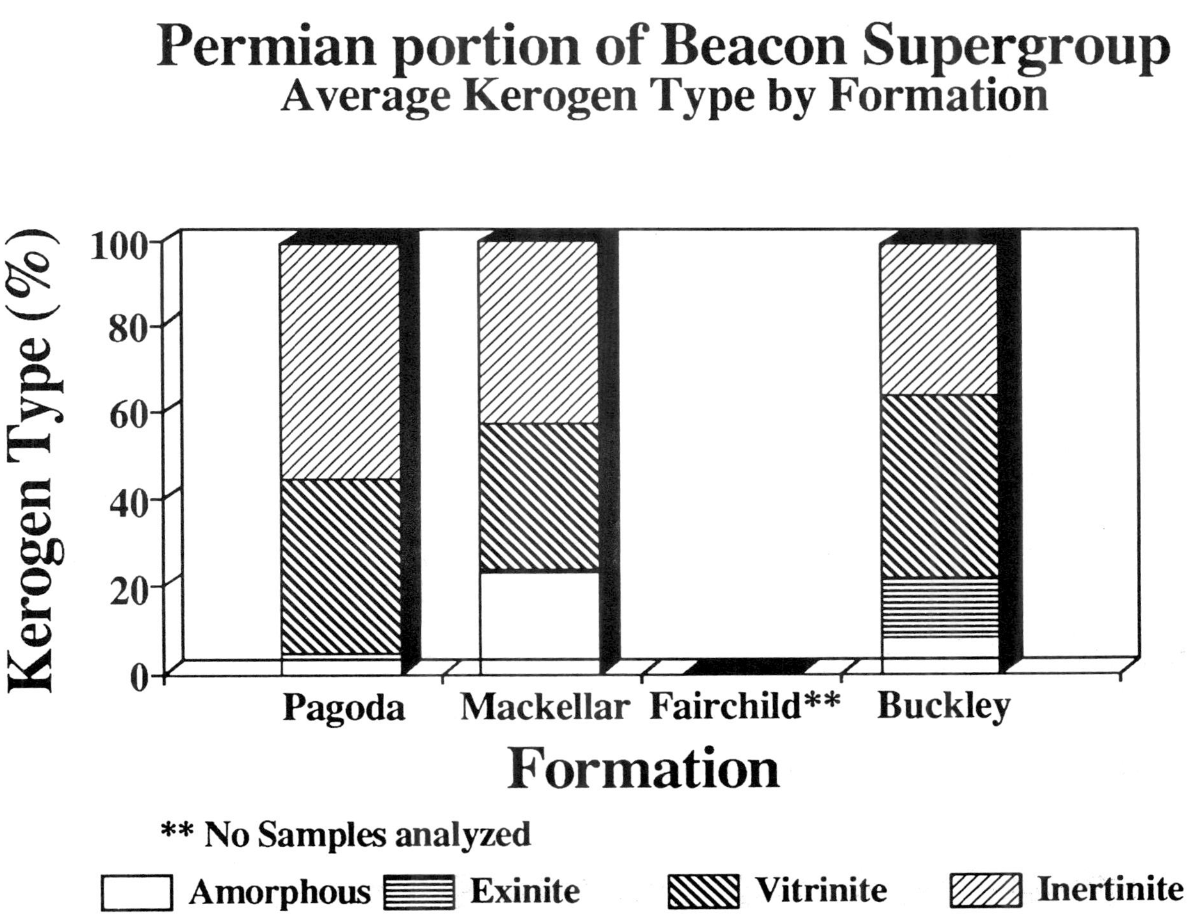

Fig. 14. Visual kerogen identification in the Permian portion of the Beacon Supergroup. The Pagoda, Mackellar, and Buckley formations are dominated by vitrinite and inertinite, which are carbon compounds derived from terrestrial sources.

Organic carbon can also be added to a depositional system by recycling organic material from older sedimentary rocks. For example, Carboniferous spores were recycled into Tertiary marine sediments of the North Sea [*Friis et al.*, 1979]. This is an unusual pathway in the global carbon cycle [*Welte*, 1969], but recycling may be important in glacial environments. Glacial processes emphasize mechanical weathering, in contrast to the chemical weathering that is dominant in many other environments. Since mechanical weathering is less likely to destroy carbon compounds, a high proportion of recycled carbon may be supplied to glacially derived sediments.

Dissolved oxygen contents, the amount of bioturbation, and the degree of bacterial degradation in sediments are all factors that influence the preservation of organic carbon. Preservation of organic carbon compounds is favored when biologic activity is low or absent, especially under anoxic bottom conditions which exclude bioturbation and prevent oxidation of the existing carbon compounds. Degradation of organic carbon is favored when biologic activity is high, the sediment/water interface is oxic, or when microbial activity destroys the carbon compounds after deposition.

In the case of the Pagoda, Mackellar, Fairchild, and Buckley formations, it is difficult to quantitatively evaluate the degree of bioturbation, oxygenation, or bacterial degradation. The important consideration for our purposes, however, is whether evidence exists for significant differences in these conditions between the four formations. Such differences could indicate that the proportion of the original organic carbon that has been preserved varies significantly through the Victoria Group. Bioturbation is rare throughout the Permian sediments, but the very limited trace fossil assemblage that is present includes vertical burrows that are similar in the Mackellar and Buckley formations [*Miller et al.*, 1990]. The presence of similar trace fossils and uniform (albeit rare) amounts of bioturbation in the Mackellar and Buckley formations indicates that the effects of bioturbation on the preservation of organic carbon do not vary significantly for these formations. Although discrete trace fossils were not observed in the Pagoda and Fairchild formations, the overall scarcity of bioturbation throughout the Permian section suggests that the preservation of organic carbon is not controlled by variations in bioturbation intensity.

In addition to the evidence provided by trace fossils, the interpretations of depositional environments presented previously provide indirect evidence about the degree of oxygenation during the deposition of the Victoria Group. Glacial Pagoda sediments were probably deposited under oxic conditions due to the turbulent nature of glacial movement and outwash. The lack of varves in Pagoda sediments may also indicate deposition under oxygenated conditions, since varves are often associated with anoxia in modern proglacial lakes. Rapid deposition of the Mackellar Formation by deltaic underflows may have enhanced water column mixing in the basin, thereby oxygenating the water column and underlying sediments. The fluvial channel environments of Fairchild and Buckley deposition were probably oxygenated, but floodplain environments of the Buckley Formation may have experienced low oxygen conditions. Ponded or peat-forming environments in the Buckley formation are the most likely example of an environment that may have been oxygen-deficient. With this exception, the influences of bioturbation and oxygen content on early postdepositional degradation of organic matter appear to have been relatively uniform through deposition of the Pagoda, Mackellar, Fairchild, and Buckley formations.

The effects of sedimentation rate on TOC content and carbon preservation cannot be easily estimated from the available data. Sedimentation rates are interpreted to have been high in the sandstones of the Mackellar, Fairchild, and Buckley formations, as evidenced by sedimentary structures such as flaser bedding and wavy bedding and the presence of thick channel deposits. Sedimentation rates in the finer-grained facies studied here are more difficult to estimate. There may have been differences in sedimentation rates between shales and siltstones in the four Permian formations, and this ultimately affects the preservation of organic carbon compounds. Differences in TOC content must be evaluated in light of all possible data, including sedimentary structures, grain size, and inferences of sedimentation rates before large-scale patterns can be properly interpreted.

The previous discussion suggests that the effects of grain size, kerogen type, amount of bioturbation, and oxygenation of the sediments on organic carbon preservation should be relatively uniform throughout the formations examined in this study. Factors that are responsible for minor variations in organic carbon preservation include differences in the original sedimentation rate and grain size effects, but these probably are not major controls on the TOC content of Beacon sediments. Two variables that have not been discussed as controls on the organic carbon content are reduction in TOC due to thermal alteration and variable input of TOC due to changes in productivity at the time of deposition. These variables will now be examined to determine which factors were the major controls on the organic carbon content of fine-grained sedimentary rocks from the Beardmore Glacier area.

Vitrinite Reflectance as an Indicator of Thermal Alteration

In the Beardmore Glacier region, Jurassic intrusions altered the sediments of the Beacon Supergroup. Diabase (dolerite) sills and dikes range from 10 cm to 200 m in thickness and have caused variable thermal alteration of the sedimentary sequence. Two major factors that control the severity of this alteration are the thickness of an intrusion and the proximity of a sample to that intrusion [*Jaeger*, 1957, 1959]. Other factors such as variable fluid migration, porosity

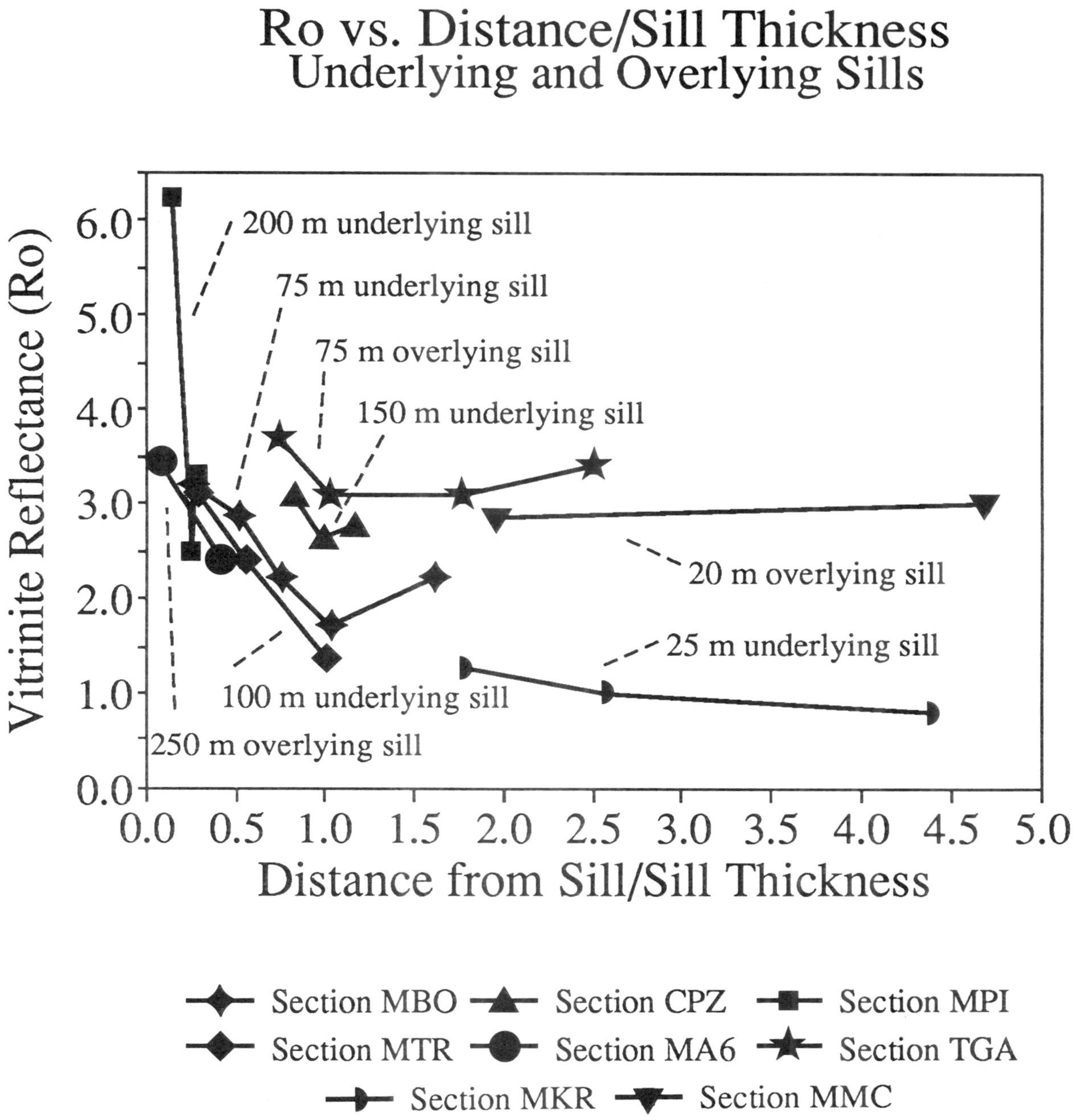

Fig. 15. Summary of data from the Permian portion of the Beacon Supergroup, comparing vitrinite reflectance to proximity to sills. Vitrinite reflectance generally decreases with increasing distance from the sill and becomes relatively uniform at less than one sill thickness. There does not appear to be a difference in the effects of underlying and overlying sills.

differences in the sediment, and compositional differences of the sediment may be important when coarse, permeable clastics are studied; but in this project these effects are held essentially constant by examining only the fine-grained clastic intervals.

Although exact values vary, several workers have suggested that the zone of severe alteration caused by an intrusion is less than twice the thickness of the intrusive body [*Bostick*, 1979; *Dow and O'Connor*, 1980]. Work by *Peters et al.* [1978] concluded that significant changes in organic matter were restricted to a much narrower zone than the above estimates would imply. Additional data from the Beardmore Glacier area can be used to estimate the size of the zone of alteration near intrusive bodies in that region.

Sill thickness data and vitrinite reflectance values from eight measured sections are shown in Figure 15. These data have been separated so that the effects of overlying and underlying sills can be evaluated independently. For each section, the distance from each sample to the nearest overlying or underlying sill was measured. The thickness of the closest sill also was recorded, and the ratio of the sill-sample distance to the sill thickness was calculated. That ratio (sill-sample distance/sill thickness) is a measure of proximity to the sill that has been normalized for sill thickness and allows data from different sections to be compared. Our data are limited to intrusions that are exposed at the surface and do not exclude the possibility that there may be hidden intrusions below the ice or outcrop surface.

Results indicate that vitrinite reflectance values are high near the sills and decrease rapidly with increasing distance from the sills (Figure 15). Samples that are farther than one sill thickness away from the sill have vitrinite reflectance values that do not vary significantly. This suggests that the effects of the nearby intrusive body are no longer important. The data points from different measured sections do not merge into a single curve beyond one sill thickness away from the sill but instead occupy a limited range of vitrinite reflectance values. This reflects the differences in background heating levels between different parts of the study area. There does not appear to be a significant difference in the effect of underlying versus overlying intrusive bodies. On the basis of these vitrinite reflectance data, the zone of significant alteration is proposed to extend one sill thickness away from a sill; samples located more than one sill thickness away from the sill are considered "unaltered."

Total Organic Carbon Contents as Indicators of Thermal Alteration

Relationships that were established using vitrinite reflectance can also be tested using TOC data. The

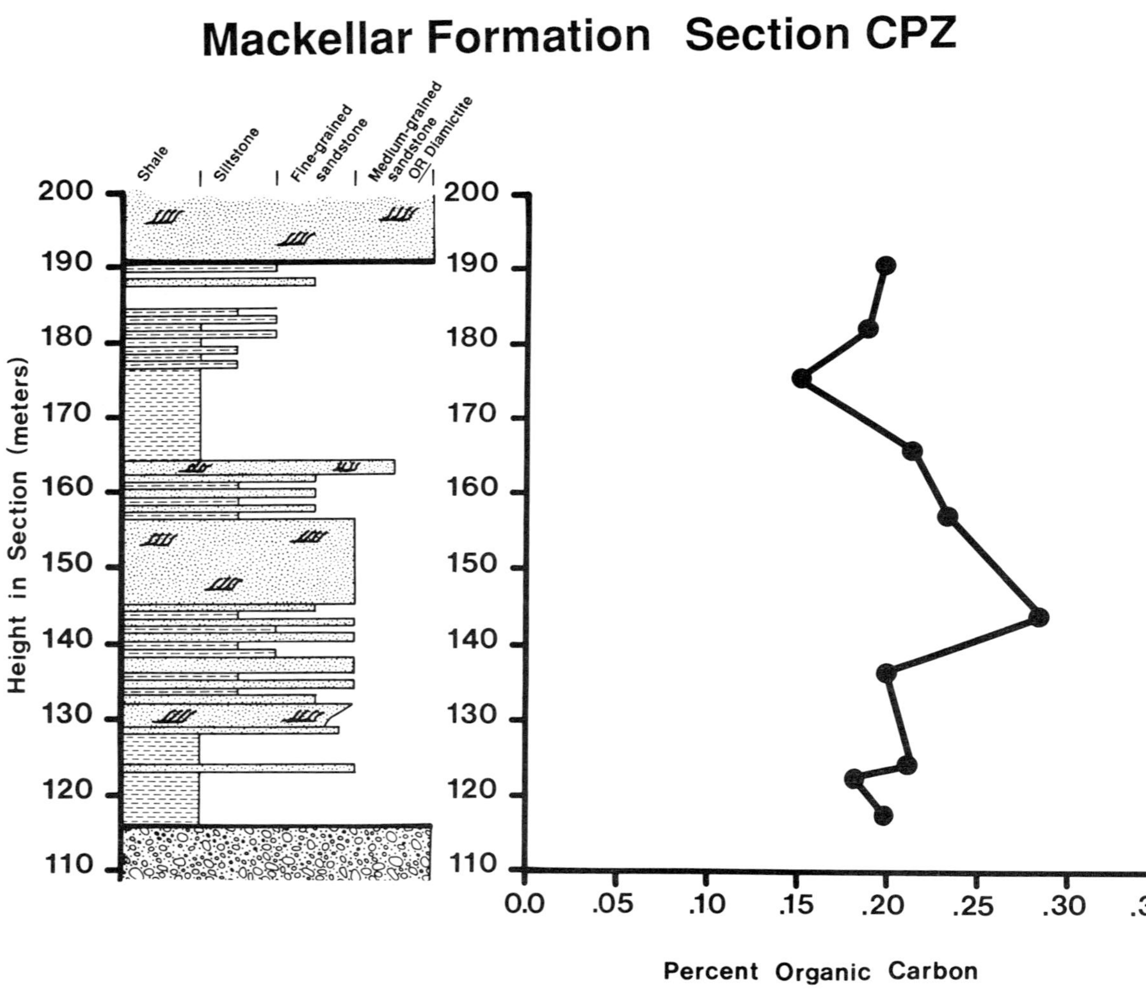

Fig. 16. Mackellar Formation, in section CPZ. The relatively constant nature of TOC content in the formation, when sills are not present, is depicted. Exclusively sampling the fine-grained beds removes variability due to grain size or composition.

Mackellar Formation is ideally suited for a study of the alteration of organic carbon because TOC contents in unaltered measured sections from the Mackellar Formation appear to be fairly uniform [*Horner*, 1990]. This allows the relationship between TOC content and distance from an intrusive body to be examined in the same manner as the vitrinite reflectance data described above.

Although TOC values in unaltered measured sections of the Mackellar Formation appear to be relatively constant, there is no geologic reason why similar internal uniformity should be true in most other clastic sedimentary deposits. The organic carbon input and preservational factors discussed above combine to cause variations in the organic carbon content of sedimentary rocks. TOC values are relatively constant in the Mackellar Formation because organic carbon input and the effects of depositional and preservational factors appear to have been relatively uniform. Minor fluctuations in TOC values within the Mackellar Formation are interpreted to reflect minor changes in depositional and preservational factors. With this in mind, major variations in TOC content that are controlled by proximity to an intrusive body are examined below.

A typical unaltered Mackellar section and the associated TOC contents are illustrated by section CPZ (Figure 16). This is an example of a Mackellar section with no exposed intrusive bodies and a relatively constant TOC content. In contrast, the Mackellar Formation in sections TGA (Figure 17) and

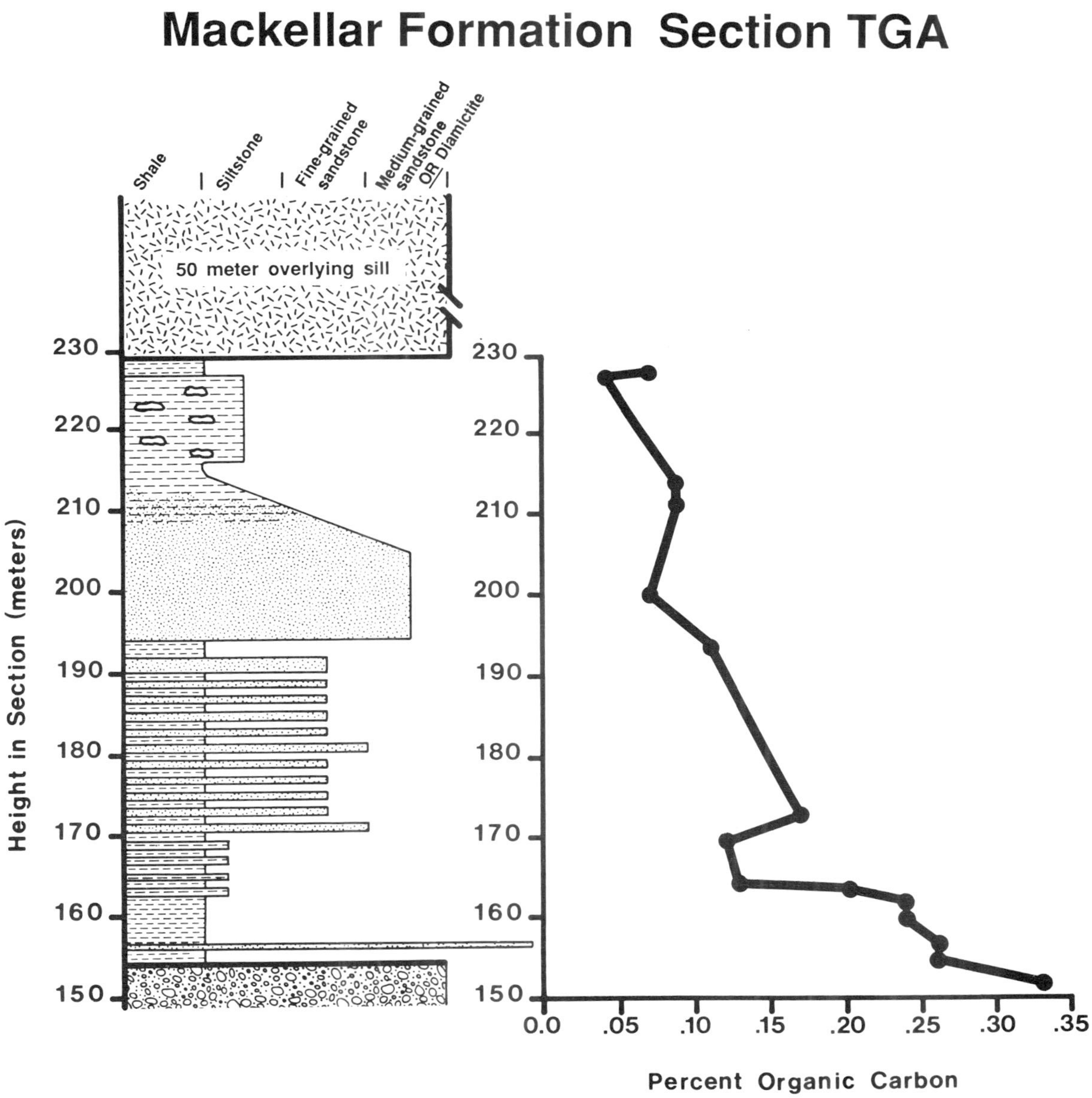

Fig. 17. Mackellar Formation, in section TGA. The effect of an overlying sill on TOC content is illustrated. An overlying sill reduces the TOC content toward the top of the section.

MMC (Figure 18) contains overlying and underlying sills, respectively. TOC content decreases near each intrusive body, with values that approach 0% TOC near the thicker sills. These field data and TOC measurements have established the basic relationship between proximity to an intrusive body and degree of alteration of the surrounding sediments. Results from several thermally altered sections are now combined to examine the relationship between TOC content and distance from an intrusive body.

Sill thickness and organic carbon distributions in six Mackellar sections are presented in Figure 19. Plots of TOC versus sill-sample distance/sill thickness indicate that the zone of carbon alteration in this study is similar to the zone of alteration reported by *Bostick* [1979] and *Peters et al.* [1978]. The most rapid decrease in organic carbon content occurs at distances of less than half the sill thickness, and most TOC profiles reach a relatively constant value at distances greater than one sill thickness (Figure 19). The TOC

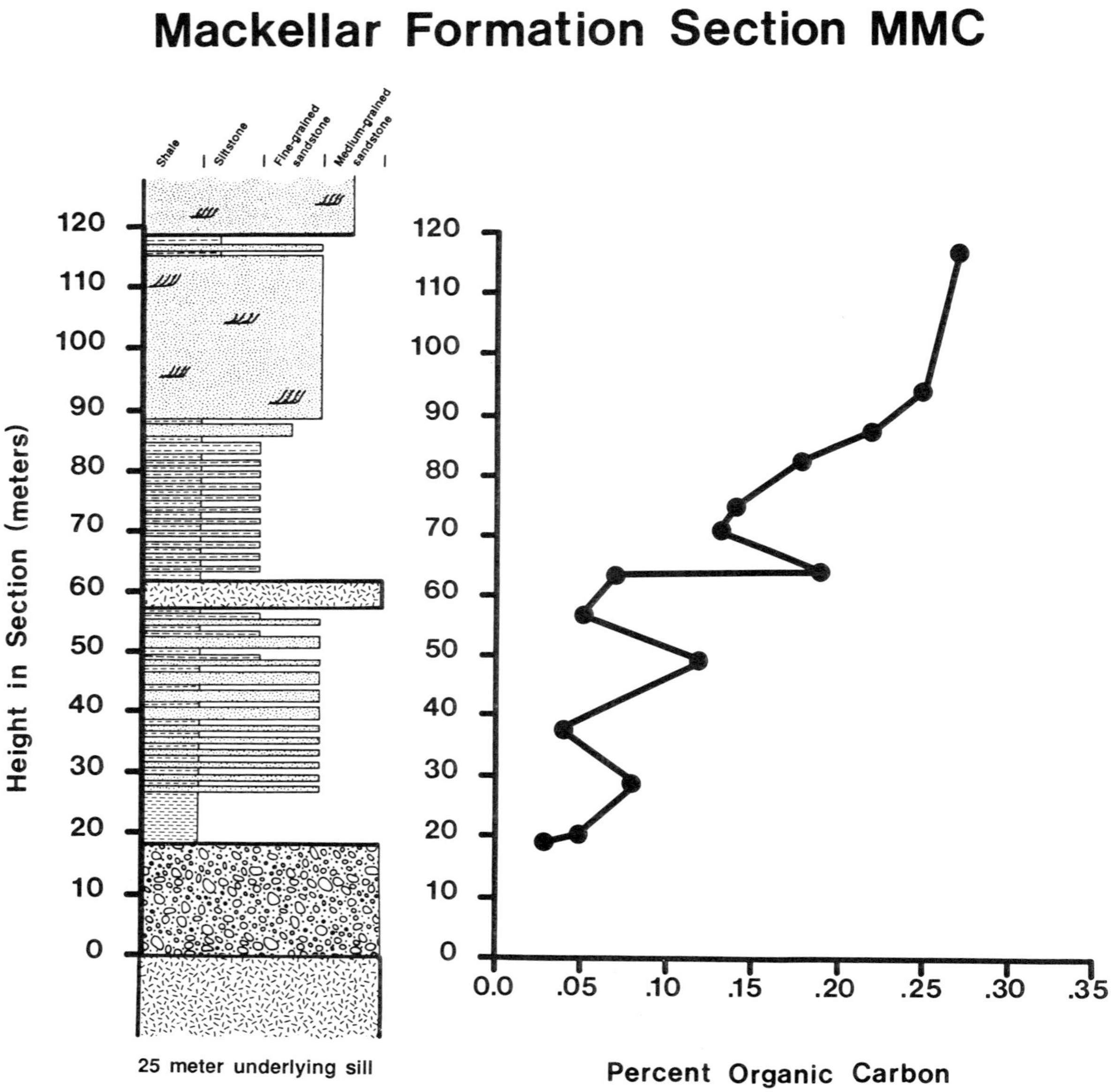

Fig. 18. Mackellar Formation, in section MMC. The effect of an underlying sill on TOC content is illustrated. An underlying sill reduces the TOC content toward the bottom of the section. The combined effects of the lower and middle sills depress the TOC content for up to 1.5 times the thickness of the lower sill. Multiple sills cause a cumulative alteration on the surrounding sediments that is greater than the sum of the alteration caused by the two sills.

content at section MKR appears to decrease over a greater thickness, but this is probably caused by a wide sampling interval near the sill. On the basis of these data, the significant effect of heating on carbon contents can be conservatively estimated to occur within one sill thickness adjacent to the sill. This provides a straightforward method for identifying altered and unaltered samples. It is interesting to note that there does not appear to be a significant difference in the behavior of underlying versus overlying intrusive bodies. Instead, natural variations in TOC content may mask any subtle differences in alteration intensity that are a result of the position of the intrusive bodies.

The effects of multiple intrusive bodies have not been considered in the preceding discussion. For example, the pattern observed in section MMC (Figure 18) is complicated by the presence of a thin (6 m thick) intermediate sill. The basal sill, which is 25 m thick, depresses the TOC contents at the bottom of

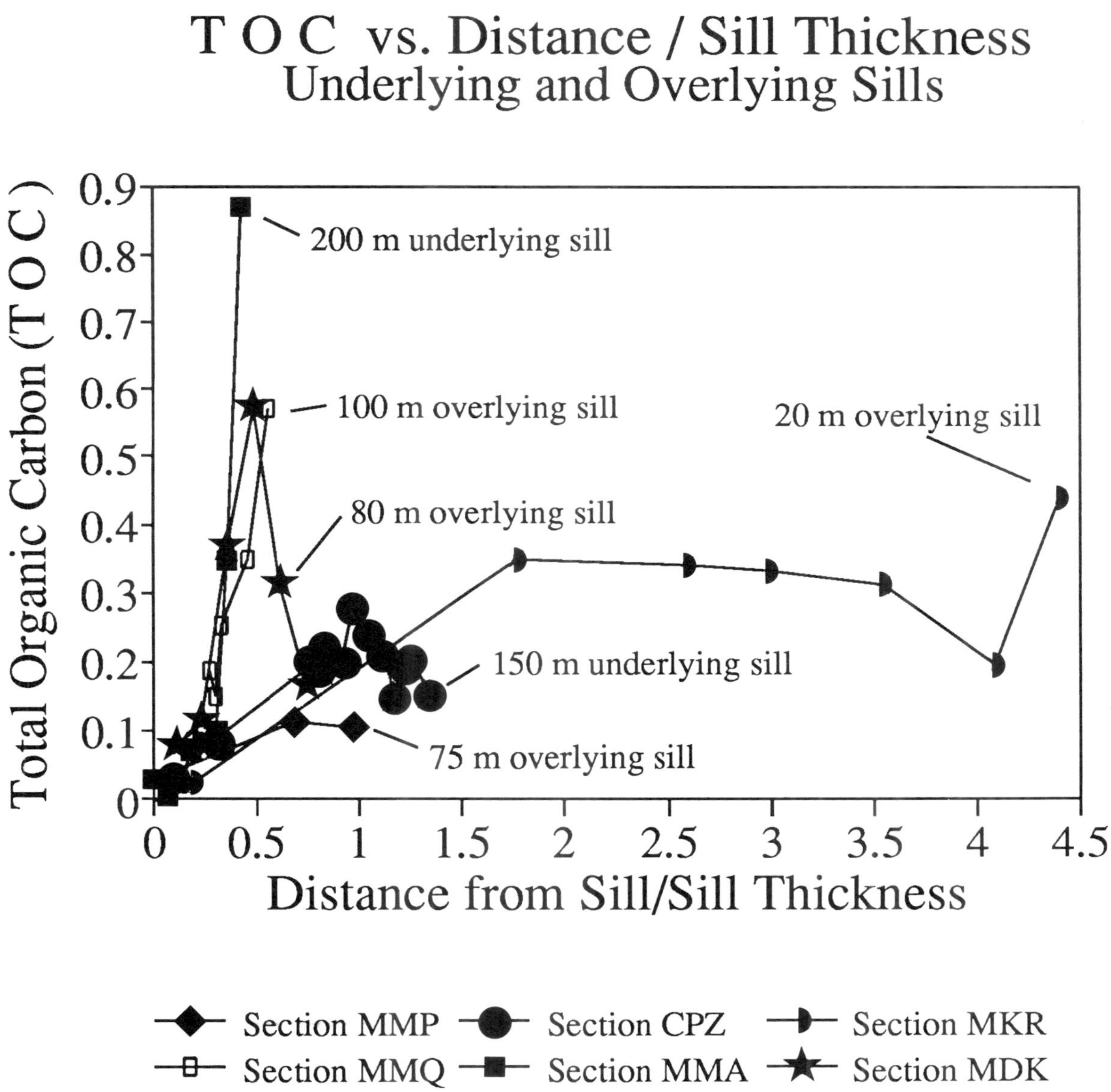

Fig. 19. Summary of data from the Mackellar Formation, comparing TOC content to proximity to a sill. Carbon content decreases rapidly at distances of less than one sill thickness. This criterion is used to identify altered versus unaltered samples. There does not appear to be a significant difference between alteration by an underlying sill and alteration by an overlying sill. Effects of multiple sills are not considered in this diagram.

the section. TOC contents then begin to rise as the effects of the basal sill become less severe, until the TOC contents are depressed again by the intermediate sill. The combined effects of these two sills reduce the TOC content in the bottom 30 to 40 m of the section, which is 1.2 to 1.6 times the thickness of the underlying intrusive body. When more than one sill is present, the zone of severe alteration is greater than the thickness of the intrusive body.

On the basis of this observation, the cumulative effects of multiple sills appear to be greater than the additive effects of the individual sills. When two sills are present, we have chosen a conservative criterion of 1.5 times the sill thickness to distinguish altered and unaltered samples. This relationship is tentative and should be used with care. *Jaeger* [1957, 1959] reached a similar conclusion with theoretical calculations by determining that preheating the country rock will dramatically increase the size of the zone of alteration. The order of the intrusive events becomes important in these calculations because early intrusion of a large sill would cause greater preheating of the country rock than early intrusion of a smaller sill. Estimates of these effects are beyond the scope of this study but highlight the complicated effects of multiple intrusions. The general prediction of severe alteration of the country rock up to a distance of one sill thickness away from the intrusive body appears to be valid for sections that contain single intrusions, but care must be used in interpreting the severity and extent of alteration when multiple intrusions are present.

Organic Carbon Contents in Least Altered Zones as Indicators of Paleoenvironments and Organic Productivity

After determining the degree of alteration of each sample, our second objective is to interpret the paleoenvironmental significance of the TOC contents of relatively unaltered samples. This was accomplished on a regional scale by comparing the mean TOC contents of the four formations studied. Since preservational effects appear to have been relatively uniform, large-scale variation in the mean TOC content is interpreted to be related to changes in productivity during Carboniferous/Permian time. TOC values are not direct measures of organic productivity, but in this study we propose that the TOC values provide a general indication of regional productivity changes during Beacon deposition. It is also important to realize that the TOC values do not stand alone as evidence for changes in depositional environment, productivity, or paleoclimate. Instead, the interpretations presented here are based on the combination of lithologic description, sedimentary structures, regional and stratigraphic relationships, and sediment compositions, which together provide a better understanding of environments at the time of deposition. Within this framework, the TOC values of relatively unaltered samples are used to provide more information about productivity within the Permian environment.

On a smaller scale, TOC contents can sometimes be used to provide a more complete description of the variety of local settings within the major depositional environments. Certain depositional conditions produced characteristic carbon contents, and identifying these conditions provides a more detailed view of Beacon deposition in the Beardmore Glacier area.

Organic carbon content, paleoenvironments, and productivity. Several widely separated localities along the paleo-Pacific margin of Antarctica have stratigraphic successions that indicate climatic warming during Carboniferous/Permian time. Sedimentologic evidence for climatic change has been described in sediments of similar age from the Ellsworth Mountains, the Pensacola Mountains, and the Ohio Range [*Elliot*, 1975; *Collinson*, 1990; *Isbell*, 1990]. These widely separated areas all contain basal glacial deposits that are overlain by coal-bearing fluvial deposits. The similarity of these successions indicates that climatic warming occurred over a large portion of Antarctica at that time.

In the Beardmore Glacier region, field-based sedimentologic and paleontologic observations provide detailed evidence of the major climatic change that occurred during Carboniferous/Permian time [*Barrett et al.*, 1986; *Miller et al.*, 1987; *Horner*, 1990; *Isbell*, 1990]. Glacial deposits of the Pagoda Formation are overlain by the shallow basinal deposits of the Mackellar Formation and, in turn, by the fluvial deposits of the Fairchild and Buckley formations. This sequence records the decay of the Carboniferous/Permian ice sheet in the Beardmore Glacier region. Within this stratigraphic framework, we propose that TOC contents provide information about general productivity levels and that these interpretations are consistent with the general patterns of climatic change recorded in the Carboniferous/Permian portion of the Victoria Group. Samples that were defined as altered on the basis of their positions relative to intrusive bodies were removed from the data set prior to this phase of analysis. TOC contents of the unaltered fine-grained samples were averaged for each formation, and the results are presented in Figure 20.

The 11 unaltered fine-grained samples from the Pagoda Formation have a mean TOC content of 0.20%, with a standard deviation of 0.12%. These low TOC values are distributed throughout several different Pagoda facies and are consistent with deposition in a low-productivity glacial setting with scarce vegetation and oxidizing conditions in the sediment. A portion of this organic carbon may be recycled material from preexisting sedimentary rocks, although the amount of recycled carbon is difficult to determine.

Thirty-nine unaltered shales and siltstones from the Mackellar Formation have a mean TOC content of 0.33%, with a standard deviation of 0.12%. The mean TOC content of the Mackellar Formation is slightly higher than that of the underlying glacial Pagoda Formation, but the respective means lie within approximately 1 standard deviation of each other.

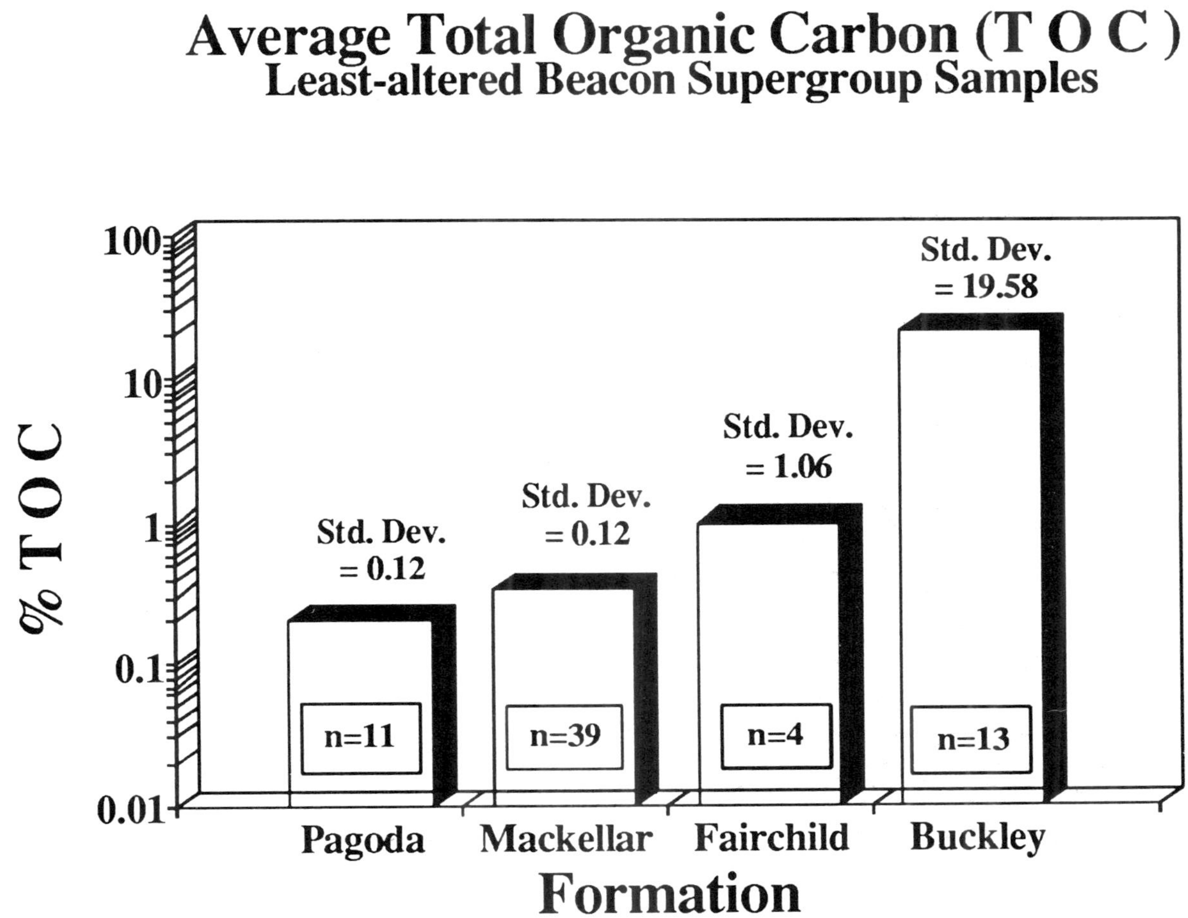

Fig. 20. Mean TOC content in the four Carboniferous/Permian formations from the Victoria Group. Carbon content and the associated organic productivity increased as the Late Paleozoic ice sheet retreated.

Both units contain predominantly land-derived organic matter (Figure 14), and both units were probably deposited under oxidizing bottom conditions. Differences in the isotope geochemistry of concretions, however, suggest that productivity levels were less similar than these mean TOC values imply.

Isotopic evidence from the concretions provides information about the sources of organic carbon in Pagoda and Mackellar sediments. The $\delta^{13}C$ values in three calcitic concretions from the Pagoda Formation average +0.9 ‰, using the Pee Dee belemnite (PDB) standard, whereas the average $\delta^{13}C$ value for six calcitic concretions from the Mackellar Formation is −9.4 ‰ (PDB) [*Lord et al.*, 1988]. The lack of ^{13}C depletion in Pagoda calcites is consistent with a lack of vegetation during Pagoda deposition, while the ^{13}C depletion in Mackellar calcites indicates that carbon was derived from oxidation of biogenic compounds [*Lord et al.*, 1988]. In combination with the previously discussed kerogen typing data (which indicates that the organic carbon in both of these formations is predominantly terrestrial), these results suggest that land vegetation was scarce and the supply of organic matter was correspondingly low while the Gondwana ice sheet was present. As the ice sheet subsequently retreated, land around the Mackellar basin became more vegetated.

In summary, productivity during Mackellar deposition was probably greater than productivity during Pagoda deposition, even though the resulting TOC values are similar. Possible explanations for the overlapping TOC values are the recycling of preexisting organic carbon during deposition of the glacial Pagoda sediments and the dilution of organic carbon by high clastic influx during deposition of the deltaic Mackellar sediments. The increase in productivity that we have inferred is probably caused by climatic warming and the postglacial dispersal of land plants in the Beardmore Glacier area.

Data from the Fairchild Formation are limited to four unaltered samples, which have a mean TOC content of 0.98% and a standard deviation of 1.06% (Figure 20). The rare fine-grained units in the Fairchild Formation are floodplain deposits, and their preservation is unusual because of the high degree of reworking in Fairchild braided stream systems [*Isbell,* 1990]. The small number of samples and an abnormally high TOC value from section MBO are responsible for the high standard deviation. When section MBO is excluded, the fine-grained Fairchild sediments have a mean TOC content of 0.38%, which is very similar to the mean TOC that was calculated for the Mackellar Formation.

The Fairchild Formation is a lateral facies equivalent to the Mackellar basinal deposits [*Miller et al.,* 1987; *Isbell,* 1990], so many of the depositional conditions that were discussed for the Mackellar Formation also apply to the Fairchild Formation. Fairchild sediments were deposited in oxygenated fluvial environments, and the organic carbon was probably supplied by terrestrial vegetation that originated in the broad, flat braidplains of the Fairchild braided stream systems. The ^{18}O isotopic data from the Mackellar Formation suggest that abundant meltwater moved through these stream systems and into the nearby basin [*Lord et al.,* 1988]. The presence of meltwater in the Fairchild streams suggests that, although the Carboniferous/Permian ice sheet had retreated, it may not have collapsed completely. The environment on the margins of the Mackellar basin was probably still variable and cool due to the effects of the decaying ice sheet.

TOC values from 13 relatively unaltered samples in the Buckley Formation average 21.64%, with a standard deviation of 19.58% (Figure 20). The high standard deviation of the Buckley TOC contents is caused by the presence of multiple facies within the larger fluvial depositional environment. The transition from Fairchild to Buckley deposition is accompanied by a large increase in mean organic carbon content, which appears to record both increased productivity and enhanced preservation.

Evidence for increased productivity, enhanced preservation, and a possible climatic change is supplied by the macroflora and the mineralogy of Buckley deposits. Plant remains are more common in Buckley sediments than in older Beacon deposits. The density of fossil tree stumps on bedding planes in the Buckley Formation indicates that forests were present [*Isbell,* 1990], and specialized *Vertebraria* root traces from the Buckley Formation show that some of these plants lived in swampy areas that bordered the Buckley stream systems [*Schopf,* 1983]. The latter observation suggests that preservation is an important factor in this increased abundance of plant remains, since preservation of organic compounds is enhanced in anoxic swampy areas. As a result, increased productivity and enhanced preservation combined to increase the TOC content in Buckley sediments.

Mineralogic evidence also supports the interpretation of climatic warming during Buckley deposition. Kaolinite generally forms as a weathering product under warm, humid conditions that favor chemical weathering. The presence of kaolinite in unaltered samples from the Buckley Formation, therefore, suggests that weathering processes during Buckley deposition were characterized by warmer temperatures and increased rainfall [*Krissek and Horner,* 1991]. This combination of paleontologic and mineralogic evidence supports the interpretation that Buckley sediments were deposited in a more moderate, humid environment.

Organic carbon content and local depositional environments. The preceding section demonstrated the usefulness of combining TOC data with field observations, paleontologic information, and mineralogic data for interpreting large-scale paleoenvironmental and productivity patterns. More localized facies relations can be clarified by mapping organic carbon abundance [*Dow,* 1977; *Schultz et al.,* 1980] or by examining organic carbon abundances within individual sections or outcrops. This approach is especially useful in the Buckley Formation, where fine-grained sediments were deposited in a variety of facies within the fluvial environment.

Facies distinctions using TOC content are of limited use in the Pagoda, Mackellar, and Fairchild formations for a variety of reasons. Fine-grained sediments from the Pagoda Formation were deposited in meltout ponds, stream overbank environments, or proglacial lakes but exhibit little variation in TOC content between these environments. In the Mackellar Formation the TOC contents of the fine-grained facies are relatively uniform. Fine-grained sediments from the Fairchild Formation were not preserved in sufficient abundance to allow any facies distinctions on the basis of TOC content.

In the Buckley Formation several facies can be distinguished using a combination of TOC values and field observations. These facies are found in floodplain settings that bordered the Buckley stream system. In some cases the TOC content alone allows the specific environment or facies to be characterized. In other cases the combination of TOC content and field relations must be used to determine the depositional environment.

Coals from the Buckley Formation have distinctive carbon values that range from 40.52 to 58.05%. Low oxygen contents and low sedimentation rates enhance the preservation of organic matter in this group of sediments. Coal beds are an example of a facies that can be characterized by TOC content, although field identification is usually the primary means for recognizing coals. High TOC values are used as supplemental evidence to confirm the environmental interpretations that were derived from field relationships, lithologies, and sedimentary structures.

A second distinct group of Buckley samples is composed of shale with rare silty shale laminae and carbon contents that range from 13.33 to 37.73% TOC. These are interpreted as sediments that were deposited in shallow-ponded settings with low oxygen conditions and sporadic clastic input. The increased importance of clastic material and the presence of rare silt laminae distinguish the pond sediments from the peat-forming deposits described above. In some cases organic-rich pond deposits grade into the sediments of the peat-forming environments, and in these cases the two facies cannot be distinguished on the basis of TOC content alone.

The third group of fine-grained Buckley samples is composed of shale or interbedded siltstone and shale, with TOC values that range from 0.19 to 12.70% TOC. This group of samples includes several different types of floodplain facies that cannot be distinguished solely on the basis of their TOC content. Instead, specific sedimentary structures or lithologic associations must be used in conjunction with the TOC content to identify facies. Paleosols are identified by the presence of plant debris and rooted horizons. Overbank deposits can be distinguished by the presence of mud cracks. Large-scale scours that are infilled with fine-grained sediment are interpreted as channel-fill deposits. In these examples, the TOC content does not distinguish the various depositional environments because the TOC contents of paleosols, overbank deposits, and channel-fill deposits overlap. TOC contents of floodplain facies and the ponded facies also overlap to some extent, but the associated sedimentary structures differentiate these major facies groupings.

The approach used in this study identifies certain facies on the basis of their organic carbon contents and provides additional understanding about the range of conditions within a depositional system when the TOC contents are combined with stratigraphic data, sedimentary features, paleontologic information, and mineralogic analysis. TOC content is most useful as an environmental indicator in the fluvial sediments of the Buckley Formation, where several distinctive facies were deposited within the braided stream setting.

SUMMARY

Field relationships, sedimentary structures, and lithologic descriptions have been used to interpret the depositional environments of four Carboniferous/Permian formations in the Beardmore Glacier region. Our studies and previous work identify the Pagoda Formation as a glacial deposit, the Mackellar Formation as a deltaic shallow basinal deposit, and the Fairchild and Buckley formations as braided stream deposits.

Several factors have the potential to affect TOC abundances in Beacon sediments. Sedimentation rates, grain size differences, types of organic matter present, recycling of preexisting carbon, bioturbation, dissolved oxygen content, thermal alteration, and variations in primary productivity have all been considered as possible causes for the observed differences in TOC content. Of these factors, thermal alteration and variations in primary productivity are identified as the variables which most often control the TOC content. The severity of thermal alteration is estimated by combining field measurements of sill thickness with laboratory measurements of vitrinite reflectance and organic carbon content. This analysis indicates that significant alteration of organic carbon is limited to a zone approximately one sill thickness away from the sill contact.

The average TOC content of relatively unaltered samples in each formation has been used in combination with other sedimentologic parameters to interpret paleoenvironments, productivity changes, and facies distributions within the Carboniferous/Permian portion of the Victoria Group. The results for the fine-grained facies of each formation are summarized below.

Pagoda Formation: Average TOC content of 0.20% with relatively uniform regional distribution.

Mackellar Formation: Average TOC content of 0.33% with relatively uniform regional distribution.

Fairchild Formation: Average TOC content of 0.98% with moderate regional variability and an abnormally high TOC content in the southern portion of the study area.

Buckley Formation: Average TOC content of 21.64%, although TOC content varies widely due to the presence of multiple facies within the fluvial depositional environment. Swamp, pond, and floodplain deposits each have a distinctive carbon content that aids in the identification of the facies.

Paleoenvironmental interpretations are made in the Permian portion of the Beacon Supergroup using evidence from field relationships, sedimentary structures, lithologic descriptions, total organic carbon analysis, kerogen typing, and carbon and oxygen isotopes. These combined data allow conclusions to be drawn that go beyond the determination of depositional environments and provide information about the productivity and climate of Antarctica during Permian time. In the Beardmore Glacier area, climatic changes during the retreat of the Carboniferous/Permian ice sheet caused an increase in the general level of terrestrial productivity, and this change is reflected in the carbon content of the sediments. The transition from the low-productivity glacial environment of the Pagoda Formation to the temperate, humid environment of the Buckley Formation is recorded clearly in these upper Paleozoic sediments.

Acknowledgments. We wish to thank the members of the Vanderbilt University and the Ohio State University field parties, whose assistance helped to shape our current thinking about these units. Appreciation is also extended to Frank Ettinsohn, Rosemary Askin, and Stephen Jacobson, who provided valuable critical reviews of the manuscript. The views expressed in this paper belong only to the authors, and other associated persons are not responsible for any errors, omissions, or differences in interpretation. Field support was provided by the U.S. Navy VXE-6 squadron. This work was funded by National Science Foundation grant DPP 84-18354. Byrd Polar Research Center contribution 765.

REFERENCES

Barrett, P. J., The post-glacial Permian and Triassic Beacon rocks in the Beardmore Glacier area, central Transantarctic Mountains, Antarctica, Ph.D. dissertation, 510 pp., Ohio State Univ., Columbus, 1968.

Barrett, P. J., G. W. Grindley, and P. N. Webb, The Beacon Supergroup of East Antarctica, in *Symp. Antarct. Geol. Solid Earth Geophys.*, edited by R. J. Adie, pp. 319–332, Sci. Comm. on Antarct. Res., Int. Union of Geol. Sci., Universitetsforlaget, Oslo, 1972.

Barrett, P. J., D. H. Elliot, and J. F. Lindsay, The Beacon Supergroup (Devonian-Triassic) and Ferrar Group (Jurassic) in the Beardmore Glacier Area, Antarctica, in *Geology of the Central Transantarctic Mountains, Antarct. Res. Ser.*, vol. 36, edited by M. D. Turner and J. F. Splettstoesser, pp. 339–428, AGU, Washington, D. C., 1986.

Bostick, N. H., Microscopic measurement of the level of catagenesis of solid organic matter in sedimentary rocks to aid exploration for petroleum and to determine former burial temperatures—A review, in *Aspects of Diagenesis, Soc. of Econ. Paleontol. and Mineral. Spec. Publ.*, no. 26, edited by P. A. Scholle and P. R. Schluger, pp. 17–43, SEPM Press, Tulsa, Okla., 1979.

Collinson, J. W., Depositional setting of Late Carboniferous to Triassic biota in the Transantarctic basin, in *Antarctic Paleobiology—Its Role in the Reconstruction of Gondwana,* edited by T. N. Taylor and E. Taylor, pp. 1–14, Springer-Verlag, New York, 1990.

Demaison, G. J., and G. T. Moore, Anoxic environments and oil source bed genesis, *Am. Assoc. Pet. Geol. Bull., 64,* 1179–1209, 1980.

Dow, W. G., Kerogen studies and geological interpretations, *J. Geochem. Explor., 7,* 79–99, 1977.

Dow, W. G., and D. I. O'Connor, Kerogen maturity and type by reflected light microscopy applied to petroleum exploration, in *How to Assess Maturation and Paleotemperature, Soc. of Econ. Paleontol. and Mineral. Short Course Notes,* no. 7, edited by F. L. Staplin et al., pp. 133–157, SEPM Press, Tulsa, Okla., 1980.

Elliot, D. H., Gondwana basins in Antarctica, in *Gondwana Geology,* edited by K. S. W. Campbell, pp. 493–536, Australian National University Press, Canberra, 1975.

Ettensohn, F. R., and T. D. Elam, Defining the nature and location of a Late Devonian–Early Mississippian pycnocline in eastern Kentucky, *Geol. Soc. Am. Bull., 96,* 1313–1321, 1985.

Friis, H., O. B. Nielsen, E. M. Friis, and B. E. Balme, Sedimentological and paleobotanical investigations of a Miocene sequence at Lavsbjerg, Central Jutland, Denmark, *Dan. Geol. Unders. Arbog,* pp. 51–67, 1979.

Frisch, R. S., Permian sandstones of the Mackellar Formation (Beacon Supergroup), central Transantarctic Mountains: Provenance and tectonic implications, M.S. thesis, 146 pp., Vanderbilt Univ., Nashville, Tenn., 1987.

Grindley, G. W., The geology of the Queen Alexandra Range, Beardmore Glacier, Ross Dependency, Antarctica; with notes on the correlation of Gondwana sequences, *N. Z. J. Geol. Geophys., 6,* 307–347, 1963.

Horner, T. C., Distribution of total organic carbon and controls on its occurrence within the Beacon Supergroup, central Transantarctic Mountains (abstract), *Geol. Soc. Am. Abstr. Programs, 22,* 15, 1990.

Horner, T. C., and L. A. Krissek, Paleogeographic interpretations using organic carbon and mineral abundance patterns in the Permian Mackellar Formation, Antarctica (abstract), *Geol. Soc. Am. Abstr. Programs, 21,* 15, 1989.

Isbell, J. L., Fluvial sedimentology and basin analysis of the Permian Fairchild and Buckley formations, Beardmore Glacier region, and the Weller coal measures, southern Victoria Land, Antarctica, Ph.D. dissertation, 347 pp., Ohio State Univ., Columbus, 1990.

Jaeger, J. C., The temperature in the neighborhood of a cooling intrusive sheet, *Am. J. Sci., 255,* 306–318, 1957.

Jaeger, J. C., Temperatures outside a cooling intrusive sheet, *Am. J. Sci., 257,* 44–54, 1959.

Jones, R. W., Organic matter near the shelf-slope boundary, in *The Shelfbreak: Critical Interface on Continental Margins, Soc. of Econ. Paleontol. and Mineral. Spec. Publ.,* no. 33, edited by D. J. Stanley and G. T. Moore, pp. 391–405, SEPM Press, Tulsa, Okla., 1983.

Krissek, L. A., and T. C. Horner, A preliminary study of REE distributions in mudrocks of the Permian Beacon Supergroup, central Transantarctic Mountains: Evidence for early development and preservation of LREE enrichment (abstract), *Am. Assoc. Pet. Geol. Bull., 72,* 208, 1988*a*.

Krissek, L. A., and T. C. Horner, Geochemical record of provenance in fine-grained Permian clastics, central Transantarctic Mountains, *Antarct. J. U.S., 23,* 19–21, 1988*b*.

Krissek, L. A., and T. C. Horner, Clay mineralogy and provenance of fine-grained Permian clastics, central Transantarctic Mountains, in *Geological Evolution of Antarctica,* edited by M. R. A. Thomson, J. A. Crane, and J. W. Thomson, pp. 209–214, Cambridge University Press, New York, 1991.

Kyle, R. A., and J. M. Schopf, Permian and Triassic palynostratigraphy of the Victoria Group, Transantarctic Mountains, in *Antarctic Geoscience,* edited by C. Craddock, pp. 649–659, University of Wisconsin Press, Madison, 1982.

Lindsay, J. F., Stratigraphy and sedimentation of the lower Beacon rocks of the Queen Alexandra, Queen Elizabeth, and Holland ranges, Antarctica, with emphasis on Paleozoic glaciation, Ph.D. dissertation, 300 pp., Ohio State Univ., Columbus, 1968.

Lindsay, J. F., Depositional environment of Paleozoic glacial rocks in the central Transantarctic Mountains, *Geol. Soc. Am. Bull., 81,* 1149–1172, 1970.

Lord, B. K., L. M. Jones, and G. Faure, Evidence for the existence of the Gondwana ice sheet in the ^{18}O depletion of carbonate rocks in the Permian formations of the Transantarctic Mountains, *Chem. Geol., 72,* 163–171, 1988.

Miller, J. M. G., Glacial advance and retreat sequences in a Permo-Carboniferous section, central Transantarctic Mountains, *Sedimentology, 36,* 419–430, 1989.

Miller, M. F., R. S. Frisch, J. W. Collinson, and W. G. Dow, Permian black shales of the central Transantarctic Mountains, in *Proceedings of the 1987 Eastern Oil Shale Symposium,* edited by R. Pettit, pp. 193–200, Kentucky Energy Cabinet Laboratory, Lexington, 1987.

Miller, M. F., J. W. Collinson, and R. S. Frisch, Depositional setting and history of a Permian post-glacial black shale: Mackellar Formation, central Transantarctic Mountains, in *Gondwana Seven Proceedings,* edited by H. Ulbrich and A. C. Rocha Campos, pp. 201–216, Instituto de Geociências da Universidade de São Paulo, São Paulo, 1990.

Peters, K. E., B. R. T. Simoneit, S. Brenner, and I. R. Kaplan, Vitrinite reflectance-temperature determinations for intruded Cretaceous black shale in the eastern Atlantic, in *Symposium on Geochemistry*, edited by D. F. Oltz, pp. 53–59, Pac. Sect., Soc. of Econ. Paleontol. and Mineral., Los Angeles, Calif., 1978.

Schopf, J. M., Forms and facies of *Vertebraria* in relation to Gondwana coal, in *Geology of the Central Transantarctic Mountains, Antarct. Res. Ser.*, vol. 36, edited by M. D. Turner and J. E. Splettstoesser, pp. 37–62, AGU, Washington, D. C., 1983.

Schultz, L. G., H. A. Tourtelot, J. R. Gill, and J. G. Boerngen, Composition and properties of the Pierre shale and equivalent rocks, northern Great Plains region, *U.S. Geol. Surv. Prof. Pap., 1064-B*, 85 pp., 1980.

Tissot, B. P., and D. H. Welte, *Petroleum Formation and Occurrence*, p. 139, Springer-Verlag, New York, 1984.

Waugh, B. J., Sedimentology and petrography of sandstones from the Permo-Carboniferous Pagoda Formation, central Transantarctic Mountains, Antarctica, M.S. thesis, 146 pp., Vanderbilt Univ., Nashville, Tenn., 1988.

Welte, D. H., Organic matter in sediments, in *Organic Geochemistry*, edited by G. Eglinton and M. T. J. Murphy, pp. 262–264, Springer-Verlag, New York, 1969.

(Received October 11, 1990;
accepted April 4, 1991.)

CONTRIBUTIONS TO ANTARCTIC RESEARCH II
ANTARCTIC RESEARCH SERIES, VOLUME 53, PAGES 67–112

THE WEST ANTARCTIC RIFT SYSTEM: A REVIEW OF GEOPHYSICAL INVESTIGATIONS

J. C. BEHRENDT,[1] W. E. LEMASURIER,[2] A. K. COOPER,[3]
F. TESSENSOHN,[4] A. TREHU,[5] AND D. DAMASKE[4]

The West Antarctic rift system extends over a 3000 × 750 km, largely ice-covered area from the Ross Sea to the Bellingshausen Sea, comparable in area to the Basin and Range and the East African rift systems. A spectacular rift-shoulder scarp along which peaks reach 4 to 5 km maximum elevation extends from northern Victoria Land–Queen Maud Mountains to the Ellsworth-Whitmore-Horlick Mountains. The rift shoulder has maximum present physiographic relief of 5 km in the Ross Embayment and 7 km in the Ellsworth Mountains–Byrd Subglacial Basin area. Various lines of evidence, no one of which is definitive, lead us to conclude that the Transantarctic Mountains part of the rift shoulder (and possibly the entire shoulder) has been rising since about 60 Ma at episodic rates of the order of 1 km/m.y. most recently since mid-Pliocene time rather than continuously at the mean rate of 100 m/m.y. We suggest (Behrendt and Cooper, 1991) a possible synergistic relation between episodic tectonism in the West Antarctic rift system and the waxing and waning of the Antarctic ice sheet (Webb, 1990) approximately coincident in time with rifting, mountain uplift, and volcanism since Oligocene (or earlier) time. The West Antarctic rift system is characterized by bimodal alkaline volcanic rocks ranging from about Oligocene or earlier to the present. These are exposed along the rift shoulder part of the Transantarctic Mountains and on the lower Amundsen-Bellingshausen flank in Marie Byrd Land and at the south end of the Antarctic Peninsula. The trend of the Jurassic tholeiites (Ferrar dolerites and Kirkpatrick basalts) marking the Jurassic Transantarctic rift crop out coincidently with the late Cenozoic volcanic rocks along the section of the Transantarctic Mountains from northern Victoria Land to the Horlick Mountains. The Cenozoic rift shoulder diverges here from the Jurassic tholeiite trend, and the tholeiites are exposed continuously (including the Dufek intrusion) along the lower-elevation (1 to 2 km) section of the Transantarctic Mountains to the Weddell Sea. Widely spaced aeromagnetic profiles in West Antarctica indicate the absence of Cenozoic volcanic rocks in the ice-covered part of the Whitmore-Ellsworth Mountains block and suggest their widespread occurrence beneath the western part of the ice sheet overlying the Byrd Subglacial Basin. A BGR-USGS aeromagnetic survey over the Ross Sea continental shelf indicates rift fabric and suggests numerous submarine volcanoes along discrete NNW trending zones. Large offset seismic profiles over the Ross Sea shelf collected by the German Antarctic Northern Victoria Land expedition V (GANOVEX V), combined with earlier USGS and other results, indicate 17–21 km thickness for the crust in the Ross Sea, which we interpret as evidence of extended continental crust. A regional positive gravity anomaly extends from the Ross Sea continental shelf throughout the subglacial area (Byrd Subglacial Basin) of the West Antarctic rift system and indicates that the crust is approximately 20 km thick rather than the 30 km reported in earlier interpretations. An approximately 200 (+50 to −150) mGal Bouguer anomaly having 4–7 mGal/km gradients where measured in places extends across the rift shoulder from northern Victoria Land possibly to the Ellsworth Mountains (where data are too sparse to determine maximum amplitude and gradient). In contrast, the maximum Bouguer gravity range is only about 130 mGal with a maximum 2 mGal/km gradient in the Pensacola Mountains area of the Transantarctic Mountains, which is easily explained by 24-km-deep Moho deepening about 8 km at a dip of 15°–20° beneath the mountains. The steepest gravity gradients across the rift shoulder require high-density (mafic or ultramafic?) rock within rifted crust as well as at least 12 km of thinner crust beneath the West Antarctic rift system in contrast to East Antarctica. Sparse land seismic data reported along the rift shoulder where velocities are greater than 7 km/s and marine data indicating lower crustal velocities about 7 km/s beneath the Ross Sea continental shelf support this interpretation. The near absence of earthquakes in the West Antarctic rift system probably results from a combination of primarily

[1]U.S. Geological Survey, Branch of Atlantic Marine Geology, Denver, Colorado 80225.
[2]Department of Geology, University of Colorado, Denver, Colorado 80217.
[3]U.S. Geological Survey, Branch of Pacific Marine Geology, Menlo Park, California 94025.
[4]Federal Institute for Geosciences and Natural Resources, Hanover, Federal Republic of Germany.
[5]College of Oceanography, Oregon State University, Corvallis, Oregon 97331.

sparse seismograph coverage, and secondarily suppression of earthquakes by the ice sheet (e.g., Johnston, 1987) and very high seismicity shortly after deglaciation in the Ross Embayment followed by abnormally low seismicity at present (e.g., Muir-Wood, 1989). Therefore we do not consider the low seismicity significant evidence against the presence of an active rift. We suggest that the origin of the West Antarctic rift system is due to a continuation of Gondwana breakup that started in the Jurassic when Africa rifted from East Antarctica (including the failed Transantarctic rift), proceeded clockwise around East Antarctica to the separation of New Zealand and the Campbell Plateau about 85–95 Ma, and has continued (with a spreading center jump) to its present location in the Ross Embayment and West Antarctica. The Byrd Subglacial Basin–Ross Embayment was probably extended greatly during late Mesozoic time, but a significant part of the total could have occurred during Cenozoic time. The Cenozoic activity of the West Antarctic rift system appears to be continuous in time with rifting in the same area that began in the late Mesozoic. The major part of the rifting, including extrusion of all of the dated alkali volcanism, took place after the separation of Antarctica from Australia and New Zealand.

INTRODUCTION

In this paper we review the geophysical research over the West Antarctic rift system over the past three decades and present new compilations including new data in the Ross Sea area. The paper is an effort to integrate these Antarctic results within the concepts of continental rifting developed for other areas during the past decade. The Gondwana breakup and the West Antarctic rift system are part of a continuously operating single system. The rifting has probably progressed clockwise around Antarctica from separation of Africa in the Jurassic to extension of continental crust in the Ross Embayment–Byrd Subglacial Basin from late Mesozoic through Cenozoic time.

One of the world's largest and least known active rift systems extends approximately 3000 km through the interior of West Antarctica, roughly parallel to the Pacific continental margin (Figure 1). First named the Cenozoic West Antarctic rift system by *LeMasurier* [1978], this largely aseismic 3000 km × 750 km area of late Cenozoic volcanism and Late Cretaceous and Cenozoic extension has also been referred to as the Ross Sea rift system [*Tessensohn and Wörner*, 1991] and the Transantarctic rift system [*Tessensohn*, 1979; *Stock*, 1989*b*] and the West Antarctic rift system by *LeMasurier* [1990*a*] and *Behrendt and Cooper* [1991]. Further complicating the nomenclature, *Schmidt and Rowley* [1986] proposed the Jurassic Transantarctic rift, to refer to an older rift system based on exposed Jurassic tholeiites along the Transantarctic Mountains extending from the Ross Sea to the Weddell Sea. Indeed, K. S. Kellogg (personal communication, 1990) has raised the question as to existence of a Jurassic rift. Kellogg noted that the only evidence for Jurassic rifting is extensive tholeiitic magmatism and the coeval timing of the initial breakup of Gondwanaland; no actual rift features (e.g., normal faults or rift basins) of Jurassic age are known. However, *Schmidt and Rowley*'s [1986] interpretation is that these features are buried beneath the ice. The Early Jurassic tholeiitic magmatic rocks (basalts, dolerite sills, and the Dufek intrusion) were intruded and extruded as part of initiation of Gondwana rifting from Africa about 184–154 Ma [*Ford and Kistler*, 1980; *Schmidt and Rowley*, 1986; *White and McKenzie*, 1989]. Tholeiitic magmatism, which generally has a shallow upper mantle origin, may only represent a failed early stage of rifting along the Ferrar-Kirkpatrick-Dufek exposures associated with the breakup of Gondwana [*Ford and Kistler*, 1980]. *Ford and Kistler* [1980] noted that Jurassic rifting along the Transantarctic Mountains ceased long before it ever reached the aulacogen stage of development.

Because the extension and rifting which we are primarily concerned with here probably began in the Late Cretaceous and continued throughout the Cenozoic, in this paper we use the term "West Antarctic rift system" to define a 3000-km-long sub–sea level bedrock valley (Figure 2) (see also *Tessensohn and Wörner* [1991]) that arcs through West Antarctica curving away from the Transantarctic Mountains between the Ross Sea and the base of the Antarctic Peninsula (Figures 1 and 2).

Recent seismic studies [*McGinnis et al.*, 1985; *Cooper et al.*, 1987*b*, 1990; *Trehu et al.*, 1989; *O'Connell et al.*, 1989] show that the crust is thinner beneath the Ross Sea part of the rift system, and older interpretations of gravity [*Bentley et al.*, 1960; *Woollard*, 1962] over the topographic trough of the Byrd Subglacial Basin reinterpreted below here indicate crustal thinning which we infer to result from late Mesozoic-Cenozoic extension. Additional gravity, aeromagnetic, and seismic studies have defined a variety of structures and anomaly patterns that seem to be characteristic of rift systems elsewhere in the world. These geophysical characteristics of the rift system are the main focus of this paper.

Because exposures in magmatic rocks correlated with the Jurassic Transantarctic rift partly overlap the West Antarctic rift system, it is important to recognize and separate the characteristics of each. The Middle Jurassic tholeiites coincide with the flank of the West Antarctic rift system along the Transantarctic Mountain front from the Pacific Ocean to within 500 km of the south pole. In the area of the Horlick Mountains, the two rift systems diverge. The Jurassic rift passes between the Whitmore and Horlick Mountains and extends into Queen Maud Land to the Weddell Sea (Figure 1) in a region of lower topography (1 to 2 km elevation) than that of the section of the Transantarctic Mountains bordering the Ross Embayment (over 4 km high) (Figure 3). In contrast,

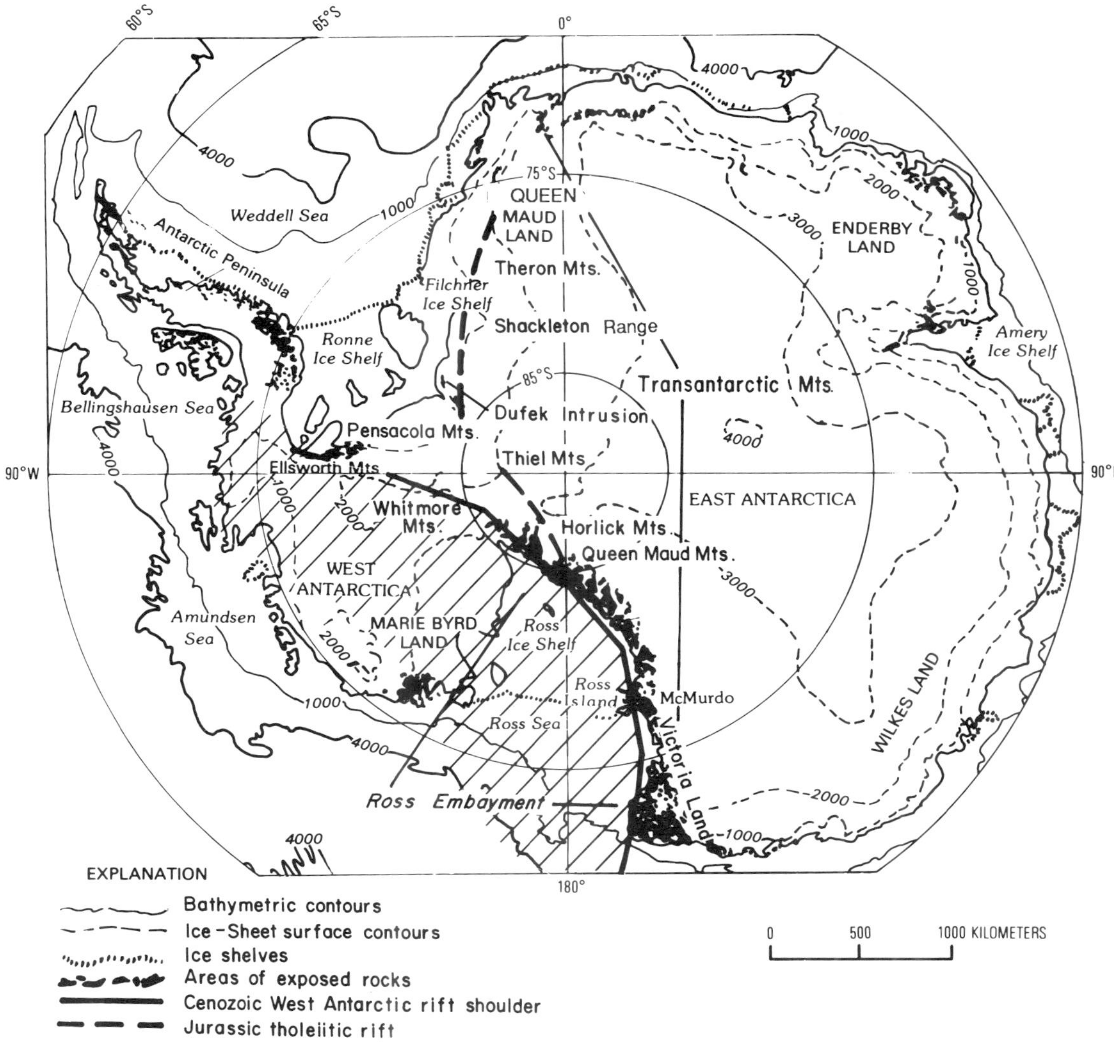

Fig. 1. Index map of Antarctica showing some of the features discussed in the text. The Transantarctic Mountains extend across the continent from Victoria Land near the Ross Sea to the Theron Mountains near the Weddell Sea and comprise the ranges shown by the heavy dashed line, as well as those bordering the Ross Embayment. The shaded area depicts the approximate location of the West Antarctic rift system. The heavy line illustrates the approximate location of the rift shoulder. To conform with convention and other publications, Figures 1, 2, 5, 12, and 13 (covering all or large regions of Antarctica) have grid north (parallel to 0° meridian) at the top. All other larger-scale maps in this paper conform to normal convention. All maps use polar stereographic projection.

the bimodal alkalic West Antarctic rift system continues beyond the Horlick Mountains and passes west of the Ellsworth-Whitmore Mountains (as indicated by aeromagnetic data discussed below) extending toward the Bellingshausen Sea. Associated late Cenozoic bimodal alkalic volcanic rocks, whose magmatic origins are inferred to be at least 100 km deep, are exposed along the west coast, in Marie Byrd Land and at the south end of the Antarctic Peninsula (Figure 2).

The distribution of igneous rocks associated with the two rifts suggests that the interior of West Antarctica has been the site of at least two episodes of lithospheric extension during the past 200 m.y., but that these episodes have only partially coincided geographically (i.e., along the Transantarctic Mountains bordering the Ross Embayment). Plate reconstructions and motions [*Norton and Sclater*, 1979; *Lawver and Scotese*, 1987; *Bradshaw*, 1989; *Stock*, 1989*a*, *b*; *Lawver et al.*, 1991] suggest that West Antarctica reached its present general configuration about 100 Ma. Similarities between crustal blocks indicate that West Antarctica is not composed of far traveled exotic terranes [*Storey et al.*, 1989].

Both geological and geophysical evidence suggests that, in addition to the late Cenozoic and Jurassic

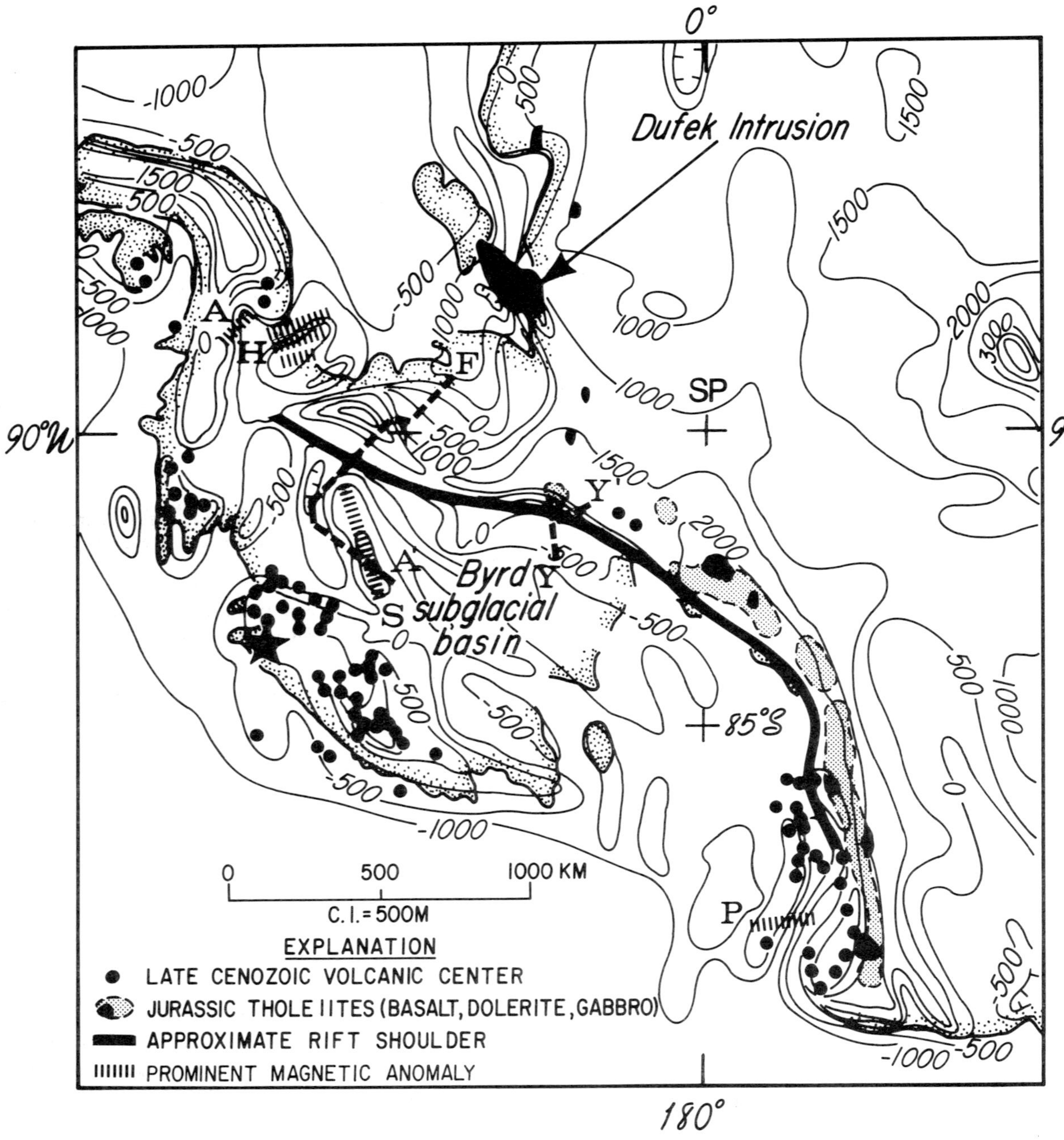

Fig. 2. Generalized isostatically adjusted bedrock elevation map after ice removal, assuming sufficient time for recovery. The contour interval is 500 m (modified from *Drewry* [1983]). All known high-amplitude linear (greater than 200 km long) magnetic anomalies greater than 1000 nT amplitude are indicated. However, coverage is only sufficient so as not to miss any magnetic anomalies in areas of the detailed aeromagnetic survey (e.g., Plates 1 and 2 and Figures 8 and 9). P, Polar 3 anomaly (Plates 1 and 2 and Figure 8); S, Sinuous Ridge anomaly [*Jankowski et al.*, 1983]; H, Haag Nunataks anomalies [*Garrett et al.*, 1987]; A, Antarctic Peninsula Traverse anomaly [*Behrendt*, 1964*b*]. The locations of late Cenozoic volcanic centers are from *González-Ferrán* [1982] and *LeMasurier and Thomson* [1990]. Locations of Jurassic tholeiites (basalt, dolerite, and gabbro) are from *Craddock et al.* [1969]. The volcanic Toney Mountain is indicated by a star (see also Figures 13 and 16). The area beneath the rift zone (elevations below sea level) is probably underlain by crust about 20 km thick, on the basis of interpretation of a Bouguer anomaly map (Figure 13), *Woollard*'s [1962] Bouguer anomaly interpretation, and seismic determinations of Moho depth beneath Ross Sea shelf (Figure 15). Locations of profiles are indicated for Figure 4 (Y—Y′) and Figure 6 (A′—F). Grid north is at the top.

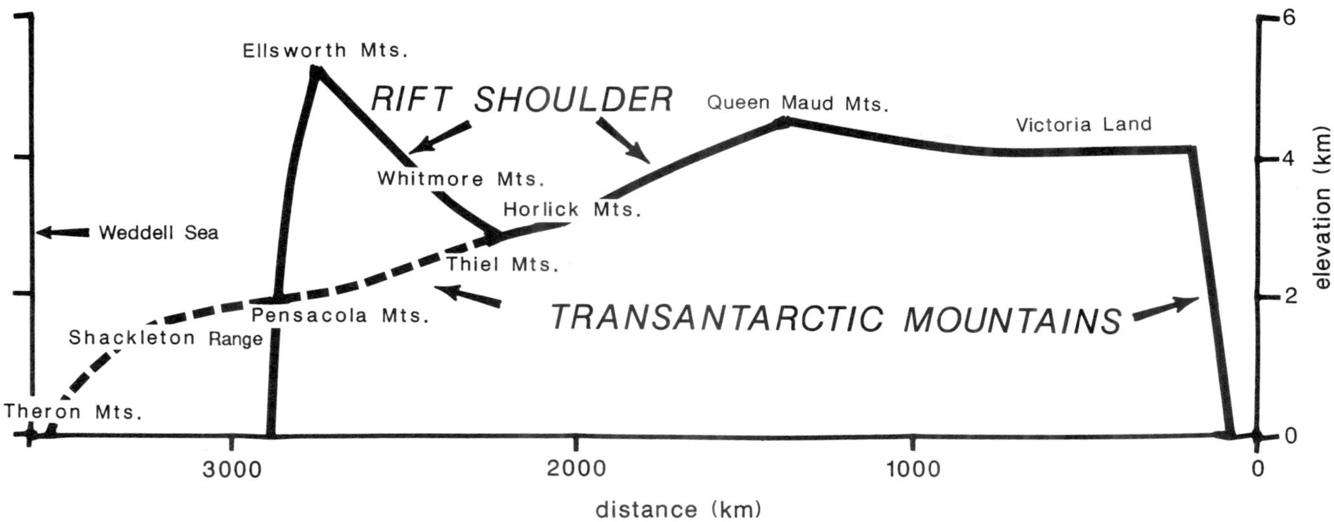

Fig. 3. Generalized topographic profile along the crest of the highest peaks parallel to and along the Cenozoic West Antarctic rift shoulder (Figures 1 and 2) (solid line) from the north coast of Victoria Land at right to the Ellsworth Mountains at left, compared with lower topography (dashed line) of highest peaks in the Transantarctic Mountains from the Horlick Mountains to the Weddell Sea. Low areas within the mountains probably are the result of glacial fluvial erosion, differential uplift, and transverse downfaulting and are not shown in these profiles.

magmatic episodes, late Mesozoic–early Cenozoic crustal extension, associated with right-lateral strike-slip displacement [*Ford*, 1972; *Storey and Nell*, 1988; *Kellogg and Rowley*, 1989], in the southern Antarctic Peninsula–Ellsworth Mountains area and [*Stock*, 1989*b*] in the Campbell Plateau area may have accompanied or immediately followed the separation of the New Zealand–Campbell Plateau block from Marie Byrd Land about 85 Ma [*Cooper et al.*, 1982; *Jankowski et al.*, 1983; *Cooper et al.*, 1987*a*; *Bradshaw*, 1989]. Although no exposed magmatic rocks of late Mesozoic age are known in the Ross Embayment–Byrd Subglacial Basin, major extension probably occurred there during this time [*Cooper et al.*, 1987*a*; *Stock*, 1989*a, b*] and continued (episodically?) in the Cenozoic. Possibly late Mesozoic volcanic rocks are buried beneath the ice or in the lower part of sedimentary basins.

The discussion of topography and the brief description of late Cenozoic volcanic rocks which follow are intended to provide an overall view of the rift. The rapid topographic uplift is discussed by *Behrendt and Cooper* [1991]. Further descriptions of the volcanic rocks are given by *LeMasurier and Rex* [1989] and *LeMasurier and Thomson* [1990].

TOPOGRAPHY

The Cenozoic West Antarctic rift shoulder, marked by spectacular nearly continuous mountain escarpments of pre-Cenozoic rocks that extend along the Transantarctic Mountains from northern Victoria Land to the Ellsworth-Whitmore-Horlick mountain front (Figures 2 and 3), has been a subject of geologic speculation since it was first described in southern Victoria Land by *Debenham* [1913]. Relief ranges from 5 km for summits of the highest peaks in the Transantarctic Mountains to the continental shelf at the edge of the Ross Sea (Figures 1 and 2), in the Ellsworth Mountains to 7 km from the summit of Mount Vinson, about 5 km above sea level, to the greatest depths of about 2.5 km in the Byrd Subglacial Basin and the Bentley subglacial trench [*Drewry*, 1983]. Subglacial erosion and graben formation have occurred simultaneously since glaciation about Oligocene time [*LeMasurier and Rex*, 1983] in the Byrd Subglacial Basin (2- to 2.5-km depths) and both have contributed to the present relief, although extension and graben formation may be as old as late Mesozoic. The sedimentary basin fill within the Byrd Subglacial Basin consists largely of moving ice, in contrast to the sedimentary rock in basins in the Ross Embayment, which accounts for the lower relief topography of the continental shelf there (Figure 2). The rugged graben-horst topography of the floor of the Byrd Subglacial Basin revealed by radar ice sounding [*Drewry*, 1983] is probably what we would expect for the Ross Embayment floor had the grounded ice sheet covered it since Oligocene time.

The Transantarctic Mountains–Ellsworth-Whitmore-Horlick flank is a well-defined rift shoulder. As discussed by *Stern and ten Brink* [1989], the part of rift shoulder bordering the best studied Ross Embayment is represented by mountains that (1) have been built in a tensional field, (2) show a lack of folding, and (3) can be accounted for by the elastic flexure of two cantilevered plates separated by a stress-free edge. In contrast, the opposing (Pacific) flank of the West Antarctic rift system, the Amundsen Sea–Bellingshausen Sea–Ellsworth Land flank (referred to hereafter as the Amundsen-Bellingshausen flank), is a highland region also with exposed outcrops in Marie Byrd Land but lacks a well-defined shoulder, resulting in overall asymmetrical topography. Rather than a single, simple linear escarpment, the Amundsen-Bellingshausen flank is marked by graben-and-horst topography [*LeMasurier and Rex,* 1983] and by pre-Cenozoic basement elevations now no more than 2500 m. The Amundsen-Bellingshausen flank is marked by much lower maximum uplifted topography than the rift shoulder, with the present highest elevations being only about 2.5 to 3 km above sea level in the exposed Marie Byrd Land basement rocks.

We examined the amount of uplift and subsidence in the West Antarctic rift system on the basis of extreme elevations (Figure 3). The steep exposed scarp that exists almost continuously along the Transantarctic Mountains from northern Victoria Land to the Horlick Mountains to the Ellsworth Mountains is interpreted to be the expression of a major normal or extensional fault that defines the boundary between the West Antarctic rift system and the rift shoulder, with a likely strike-slip component [*Ford,* 1972; *Storey and Nell,* 1988; *Kellogg and Rowley,* 1989]. Figure 4 shows a subglacial view of the steep scarp and is suggestive of a very youthful topography [*Behrendt and Cooper,* 1991].

The range in extreme topography seen in Figure 3 along the rift shoulder is probably due to erosion as well as uplift and downfaulting. Based on youthful morphology, age of volcanism, tilted Beacon Supergroup rock, elevation, and maximum relief, we interpret the main cause of uplift along the Cenozoic West Antarctic rift shoulder to be late Cenozoic tectonism associated with rifting. The Cenozoic uplift can be modeled for a cantilevered lithospheric plate heated at the free edge [*Stern and ten Brink,* 1989].

Maximum elevations (Figure 3) vary from about 4 km in Victoria Land and the Queen Maud Mountains, dropping to about 3 km in the Horlick and Whitmore Mountains, but rise to 5 km in the Ellsworth Mountains (Figure 1). The abrupt drop at the north end of the Ellsworth Mountains separates these mountains from the southern Antarctic Peninsula and was interpreted as a major tectonic break by *Behrendt* [1964*b*] and as a rift by *Doake et al.* [1983] and *Maslanyj and Storey* [1990]. The drop in elevation in the Horlick and Whitmore Mountains may have resulted from either a greater rate of erosion or differential uplift. In contrast, elevations along the Transantarctic Mountains are lower toward the Weddell Sea (see dashed-line profile in Figures 1 and 3). The maximum elevations range from about 2 km in the Thiel and Pensacola Mountains (including the Dufek intrusion) and drop to about 1.5 km in the Shackleton Range and 1 km in the Theron Mountains. The low scarp bordering these ranges defines the boundary of the 1.7-km-deep Thiel-Crary trough beneath the Filchner Ice Shelf [*Behrendt et al.,* 1974]. We emphasize that the lower topography at the Weddell Sea end of the Transantarctic Mountain chain is not a part of the late Cenozoic rift. The age of uplift in this area is not known but appears to be significantly older than that of the West Antarctic rift system, partly based on the fact that the Cenozoic sedimentary rocks within the Thiel-Crary trough are tectonically undeformed [*Haugland,* 1982].

Uplift of Rift Shoulder

The steep scarp marking the rift shoulder suggests a youthful topography. Although present rates of weathering may be low in the absence of water and although dry-based glacial erosion may be minimal, repeated periods of Cenozoic glaciation and deglaciation [*Webb,* 1990] would probably have been accompanied by very high rates of fluvial and wet-based glacial erosion. We interpret that the main cause of uplift, along the rift shoulder (Figures 1 and 2), is late Cenozoic tectonism, probably as modeled by *Stern and ten Brink* [1989], for a flexed continental lithospheric plate heated at the free edge (mountain front).

The variation of the high topography seen in Figure 3 along the rift shoulder is partly caused by erosion and by differential uplift on transverse faults. *Denton et al.* [1984] refer to possible separate histories of individual fault blocks. They note that uneroded subaerial volcanic cones dated 4.2 and 3.5 Ma in Wright Valley (and other evidence) indicate that tectonic uplift is limited to about 300 m since the last glacial overriding. *Webb and Wrenn* [1982], *Wrenn and Webb* [1982], and *Ishman and Webb* [1988] also show evidence that indicates about 500 m of uplift occurred in the Taylor Valley area starting about 3 Ma. Although the maximum elevation in the Transantarctic Mountains in the McMurdo area (e.g., Mount Lister) is 4 km, about 20 km to the north, maximum elevations on ridges bordering the Taylor and Wright valleys are only about 2 km, suggesting to us differential uplift on transverse faults along the axis of the mountains in this area. *Wrenn and Webb* [1982] discussed differential movement in more detail in the dry valleys area. Transverse faulting along rift shoulders is common and is apparent [*Cooper et al.,* 1991] along the Transantarctic Mountains bordering the Ross Embayment. Transverse vertical faulting resulting in differential uplift is apparent between McMurdo and the Shackleton Glacier (about 85°S, 178°W, not shown) [*Katz,* 1982; D. Elliott, personal communication, 1990]. Aerial photography taken at an altitude of 12.5-km elevation along the Transantarctic Mountains north of McMurdo suggests faults transverse to the strike of the Transantarctic Mountains here. We interpret this as evidence of differential uplift (or subsidence) possibly along unrecognized transverse faults.

Fitzgerald [1989], using fission track dates from the relatively low dry valley area north of Mount

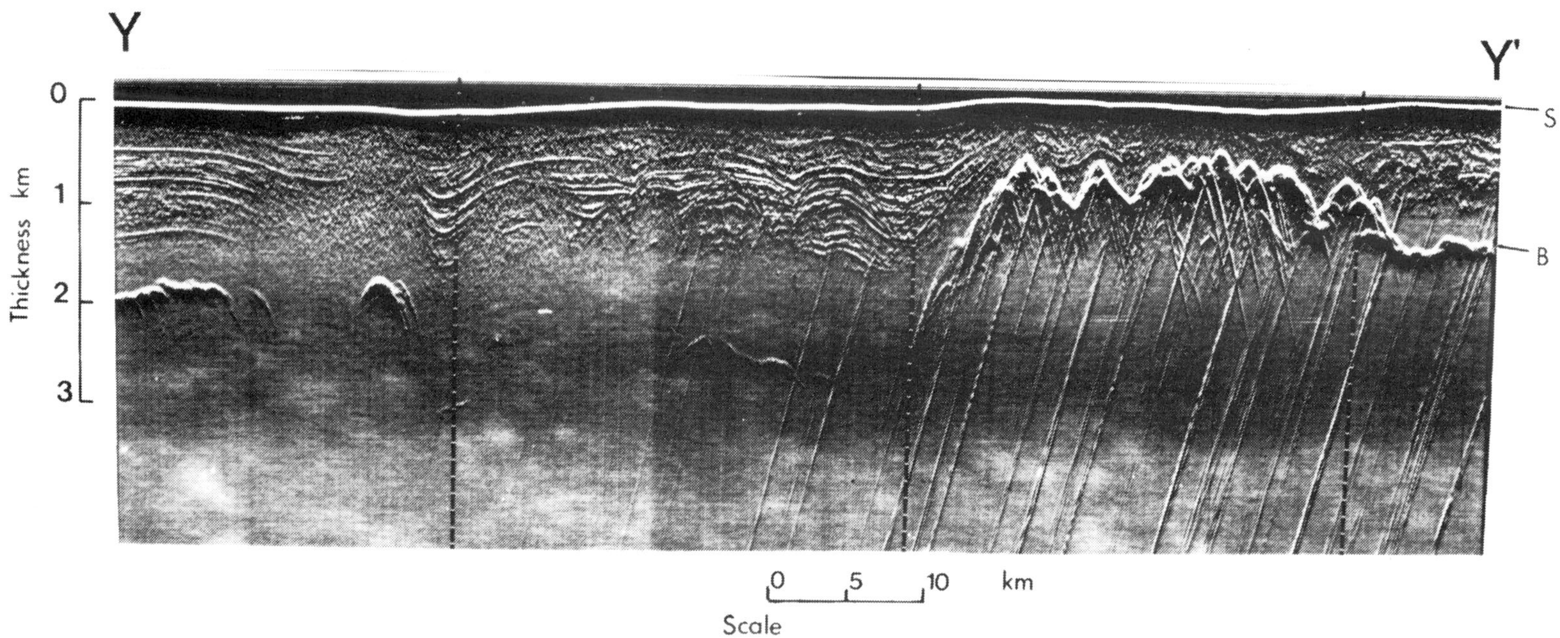

Fig. 4. Example of a radar ice-sounding profile Y—Y′ across the ice-covered part of the rift shoulder, from *Jankowski et al.* [1983]. The location is indicated in Figure 2. S, snow surface reflection; B, bedrock reflection.

Lister, interpreted 5 to 6 km of uplift in the Transantarctic Mountains of southern Victoria Land. Uplift began about 60 Ma and continued at an average calculated rate of about 100 m/m.y. since that time. These estimates are similar to those summarized by *Tingey* [1985]. The extreme amount of uplift which *Fitzgerald* [1989] inferred in northern Victoria Land is 10 km. Seismic reflection data from the Ross Sea adjacent to the Transantarctic Mountains show several distinct angular unconformities along the western edge of the Victoria Land basin, which we interpret as evidence for episodic uplift. Therefore, we infer (as did *Behrendt and Cooper* [1991]) that the uplift of the Transantarctic Mountains and entire rift shoulder was episodic and has probably been an order of magnitude faster, at times including the present, than the mean rate. Data from other similar tectonically active rift areas in the world suggest that uplift rates as great as 1 to 2 km/m.y. are not uncommon as discussed below, and thus we believe these high rates are reasonable for the rift shoulder.

Using an Occam's Razor or "principle of least astonishment" approach, various lines of evidence, none of which are independently conclusive, led *Behrendt and Cooper* [1991] to interpret that the Transantarctic Mountains part of the rift shoulder (and probably the entire shoulder) has been rising at a rate of approximately 1 km/m.y. during the latest episode, probably since early or middle Pliocene time. The following is evidence [*Behrendt and Cooper*, 1991] for this order of magnitude of uplift, no single item of which is definitive or which cannot be explained by other mechanisms, as several of the authors cited have done.

1. The youthful-appearing rift shoulder has about the highest elevation (4 to 5 km, Figure 2) of any suspected rift mountains on Earth.

2. The 5–7 km relief on the shoulder (Figure 2) is the greatest of any rift mountains, even if the rift fill (ice) were compressed to the density of sedimentary rock, and that elevation contrasted with shoulder elevations.

3. Holocene fault scarps with about 10-m displacement (discussed in a later section) are interpreted from high-resolution marine (Ross Sea) reflection profiles.

4. Normal faults displace late Pliocene moraines, with as much as 300-m vertical displacement along the Transantarctic Mountains bordering the Ross Embayment [*McKelvey et al.*, 1991].

5. Unconformities in the Victoria Land basin [*Cooper et al.*, 1987*a*] in conjunction with the 100 m/m.y. average uplift rate since 60 Ma [*Fitzgerald*, 1989] require episodic uplift. We do not know of any active rift shoulder with a constant uplift rate for 60 m.y.

6. Glacial striae are found on peaks in the Transantarctic Mountains 600–1000 m above present ice levels [e.g., *Höfle*, 1989; *McKelvey et al.*, 1991]. Although these striae possibly indicate higher ice sheet elevations in the past [*Höfle*, 1989; *Prentice et al.*, 1986], we think that the striae are more likely due to tectonic uplift of the peaks.

7. *Nothofagus* (Pliocene beech leaf) occurs in situ in Pliocene moraines at high elevations in the Transantarctic Mountains [*Webb and Harwood*, 1987]. This presence is consistent [*Mercer*, 1987] with, but does not necessarily require, a 1 km/m.y. uplift rate (Harwood et al., unpublished data).

8. Late Pliocene marine microfossils in the Sirius Formation have been glacially transported from the Wilkes basin by the East Antarctic ice sheet to elevations between 1750 and 2500 m in the Transantarctic Mountains. These microfossil locations can be more easily explained if the mountains were much lower at 3 Ma than at present [*Webb et al.*, 1984].

9. Neotectonic topography of other rifts has uplift rates that are similar or higher (e.g., East Africa, 2 and 8 km/m.y. [see *Ebinger*, 1989]; western United States (Wasatch Mountains), 2 to 2.5 km/m.y. [see *Naeser et al.*, 1983]; and other areas [see *Ruddiman et al.*, 1989; *Parry and Bruhn*, 1987; *Lajoie*, 1986]).

Behrendt and Cooper [1991] considered age limits on the latest tectonic episode. The present Victoria Land–Ellsworth Mountains scarp could have been built in 4 to 5 m.y., assuming that all uplift was accommodated at about 1 km/m.y. We know however [e.g., *Fitzgerald*, 1989; *Barrett*, 1989] that uplift started much earlier. The age of the uplift of the Pensacola-Shackleton-Theron part of the Transantarctic Mountains is unknown, but the faulting of the Transantarctic Mountains from which offsets the Jurassic Dufek intrusion is younger than its emplacement [*Behrendt et al.*, 1974]. Therefore, alternatively assuming that the 2-km elevation of the Pensacola Mountains (Figure 3) is equal to that of the entire rift shoulder (from Victoria Land to the Ellsworth Mountains) prior to the latest (i.e., since Pliocene) episode of uplift, only 2 to 3 m.y. are needed for the rift shoulder to reach the present highest elevation (4 to 5 km). Estimates from *Harwood* [1986], *McKelvey et al.* [1991], *Webb and Harwood* [1987], and Harwood et al. (unpublished data) provide considerably better timing of this uplift in the Queen Maud Mountains area, where the start of uplift must postdate the deposition of the Sirius Group and the Pliocene *Nothofagus* between 2 and 3 Ma (Harwood et al., unpublished data).

Earlier episodes of uplift and erosion also occurred, as suggested by the maximum uplift of 10 km in northern Victoria Land [*Fitzgerald*, 1989] and interpretations by *Barrett* [1989] that the Transantarctic Mountains had reached half their present elevation by early Oligocene time. Uplift history along the entire 300-m rift shoulder was episodic and complex with transverse faulting and differential uplift.

Amundsen-Bellingshausen Flank

The Amundsen Sea flank of the rift system is marked by lower maximum elevations, compared to the Transantarctic-Horlick-Ellsworth Mountains flank, for uplifted pre-Cenozoic basement. The magnitude and timing of the vertical displacements are also better constrained. The most conspicuous topographic features are the high volcanic peaks, which occur both in isolated and rectilinear chains, and the flat-topped coastal nunataks, which are distributed in a manner that simulates basin-and-range topography [*LeMasurier and Rex*, 1983]. The key structural datum in this region is a late Cretaceous–early Tertiary erosion surface of very low relief that bevels pre-Cenozoic granitic and metamorphic basement rocks throughout the mountainous uplifted area of Marie Byrd Land and Ellsworth Land [*Craddock et al.*, 1964; *LeMasurier and Rex*, 1982]. The erosion surface is exposed at different levels in different nunataks, at elevations ranging from 500 m to 2700 m, and is typically overlain by a thin sequence of basaltic flows and hydroclastic deposits. Dating of these volcanic rocks has revealed a systematic pattern of increasing age on successively higher remnants of the erosion surface. The displacements of the erosion surface seem to be the result of block faulting, as there is no evidence anywhere for multiple erosional surfaces, and the difference in elevation of the surface from one nunatak to another is more than 40 times the measurable erosional relief on the surface. The systematic pattern of increasing volcanic ages with higher elevations has been interpreted to represent sequential uplift of individual fault blocks, where the initiation of uplift was accompanied by volcanic eruptions [*LeMasurier and Rex*, 1983]. These relationships seem, at the very least, to represent 2.2 km of differential vertical uplift within the past 25 m.y. (i.e., 120 m/m.y.). Arguments for a higher episodic uplift rate for the rift shoulder probably apply here as well.

The regional extent of the erosion surface provides a basis for inferring subsidence when it is considered together with geophysical and chronologic data. The seismic profile at late Miocene (10.1 Ma [*LeMasurier*, 1990*a*]) Toney Mountain (75°45′S, 117°00′W) (Figure 2) described in a later section provides evidence that magnetic anomaly–producing rocks with seismic velocities similar to the volcanic rocks lie 3 km below sea level. From this we infer that the volcanic section extends to 3 km below sea level where it is in contact with basement rock inferred from seismic velocities. The erosion surface interpreted at depth is exposed only 80 km north of Toney Mountain. Hence, assuming that the erosion surface formed at or above sea level, it appears that there has been at least 3 km of posterosion surface subsidence beneath Toney Mountain.

The most significant aspect of topography of the exposed rock area in Marie Byrd Land is that it appears to record up to 2.7 km of horst uplift and up to 3 km of graben subsidence and is suggestive of an area within a rift rather than a rift shoulder. In contrast, the available data suggest that Cenozoic deformation on the Transantarctic Mountains flank was dominantly, perhaps exclusively, uplift and was largely subsidence associated with extension along the axial part of the rift system. Possibly this asymmetry in uplift characteristics between the rift shoulder and the Amundsen-Bellingshausen flank is caused by the very rigid lithosphere underlying East Antarctica [*Stern and ten Brink*, 1989] contrasted with (probably less rigid) oceanic lithosphere offshore. Geophysical studies indicate multiple periods of subsidence in grabens in Ross Sea basins since as early as late Mesozoic [*Cooper et al.*, 1987*a*], which is probably characteristic of the ice-covered low areas in the Ross Embayment–Byrd Subglacial Basin.

CENOZOIC VOLCANIC ROCKS

The large late Cenozoic volcanoes along the west coast of the Ross Sea and those along the Pacific coast of Marie Byrd Land [*LeMasurier*, 1990*a*, *b*] are, along with the rift shoulder, the most conspicuous features of the West Antarctic rift system. The highly alkaline, bimodal character of the volcanic rocks suggested even to early workers [*Hamilton and Boudette*, 1962] that volcanism in this region was rift related. However, the highly asymmetric distribution of volcanic fields and the ice cover obscured both the geometry and scale of the rift system. Quite likely the bulk of the volcanic activity, however, occurred beneath the ice and sea-covered low-elevation parts of the rift proper rather than along its flanks. Using gravity, seismic and magnetic patterns, and the reconnaissance mapping of subglacial topography by radar–echo-sounding techniques (Figure 2) [*Drewry*, 1983], it is possible to present an integrated view of the distribution of exposed volcanic rocks in relation to the still somewhat tentative boundaries of the rift system. Chronologic patterns of volcanism and the specific geochemical characteristics of the rocks add still more to the overall picture of the structure and evolution of the West Antarctic rift system. *Hole and LeMasurier* [1990] reported a roughly threefold increase in the age range of volcanic activity in the West Antarctic rift system compared with late Cenozoic alkaline basalts along the Antarctic Peninsula. They also estimated 3 orders of magnitude increase in the volume of eruptive products in Marie Byrd Land, which they considered evidence of a mantle plume.

Distribution

Cenozoic volcanic rocks are exposed discontinuously and asymmetrically along both flanks of the rift system from the Ross Sea to the Bellingshausen Sea (Figure 2) and southern Antarctic Peninsula. It can be seen from Figure 2 that these exposures are arrayed roughly parallel to the deep subglacial topographic trough that forms the axial part of the rift. The possible occurrence of subglacial volcanoes in this axial region can only be inferred from geophysical studies, and the least ambiguous possibilities are those along the floor of the Ross Sea, described below. The largest volcanoes and the only known occurrences of felsic rock related to the West Antarctic rift system are found along the Ross Sea coastal segment of the Transantarctic Mountains (between about 71° and 79°S) and along the Amundsen Sea coast of Marie Byrd Land between about 110° and 140°W (Figures 1 and 2). The major volcanoes of these two regions are typically shieldlike in form and structure (although Mount Erebus and Mount Melbourne are stratovolcanoes), display between 1 and 3 km of constructional relief and are between 10 and 50 km in base diameter. Basaltic rocks occur in these shield volcanoes, in thick laval (~2 km) and hydroclastic sequences of undetermined structure, and also in numerous minor occurrences of flows, cinder cones, and tuff cones. Felsic rocks (trachyte, phonolite, pantellerite, and comendite) occur either as summit sections of basalt shield volcanoes or as the major component of felsic shield volcanoes. The latter contain very small proportions of pryoclastic rock compared to typical parts of the world [*LeMasurier*, 1990*a*]. Intermediate rocks (benmoreite and mugearite) are usually associated with the felsic rocks in the West Antarctic rift system and are relatively uncommon. In Marie Byrd Land it has been estimated that basaltic rocks make up between 70% and 90% of the total volume of Cenozoic volcanic rock [*LeMasurier*, 1990*a*, *b*]; the Ross Sea provinces are similar in these characteristics [*Kyle*, 1990]. The basaltic rocks in general produce the high-amplitude magnetic anomalies discussed below.

In contrast, volcanoes at the Bellingshausen Sea end of the rift system are relatively small, widely scattered, and entirely basaltic. This area includes all of the volcanoes in Ellsworth Land and the southern end of the Antarctic Peninsula (Figure 1). These generalizations are not evident from Figure 2, which shows only schematic representations of known Cenozoic volcanic localities, irrespective of their size or composition. The relevance of these distribution patterns to geophysical interpretations is that one can be sure that Cenozoic volcanic rocks are a dominant rock type between about 71° and 79°S, along the west coast of the Ross Sea, and between 110° and 140°W, along the Pacific coast of Marie Byrd Land, and thus they provide "ground truth" for comparison with areas dominated by igneous rocks that are not related to the rift system.

Geochemical Characteristics

Volcanic rocks associated with the West Antarctic rift system are an alkaline bimodal assemblage, all of which belong to the sodic series of alkaline rocks [*LeBas et al.*, 1986]. They are similar in many ways to rocks in the eastern rift of the East African rift system, but the highly potassic rocks and carbonatites of the western rift in East Africa have not been found anywhere in the West Antarctic rift system.

The basaltic rocks are remarkably similar to one another throughout the rift system and indeed throughout West Antarctica as a whole [*Hole and LeMasurier*, 1990]. They are typically nepheline-normative and olivine-phyric and contain spinel lherzolite and other mantle-derived nodules in a few widely scattered localities (e.g., transitional alkali basalts and basinites in the Mount Melbourne volcanic field [*Wörner et al.*, 1989]).

Where complete data sets are available, they reveal especially narrow ranges in La/Nb (0.6–1.4), $^{87}Sr/^{86}Sr$ (0.70258–0.70387), and $^{143}Nd/^{144}Nd$ (0.51285–0.51307), indicating that basalts throughout the rift system were derived from a depleted mantle source region (i.e., within the asthenosphere) with no preceptible contribution from lithospheric mantle or crustal sources [*Stuckless and Ericksen*, 1976; *Futa and LeMasurier*, 1983; *Wörner et al.*, 1989; *Hole and LeMasurier*, 1990]. Some heterogeneity within this source region must be at least partly responsible for variations in the Zr/Nb and other high field strength element ratios among closely associated basalt localities [*Hole*, 1988; *LeMasurier*, 1990*a*, *b*], and variable but low degrees of partial melting are believed to be responsible for other geochemical variations, most notably in Nb/Y and Ta/Yb ratios [*Hole*, 1988].

None of the variations described above serves to distinguish basalts related to the West Antarctic rift system from non-rift-related postsubduction basalts from the Antarctic Peninsula. For this purpose, the best discriminators are relative and absolute abundances of K, Ba, and U, and a variety of ratios involving these elements. Rift-related basalts from the Ross Sea region, Marie Byrd Land, and western Ellsworth Land all have K/Ba ratios of <50, whereas Antarctic Peninsula basalts have K/Ba ratios of >50 [*Hole and LeMasurier*, 1990]. Furthermore, the geochemical characteristics of the rift-related basalts are indistinguishable from those of oceanic island basalts, and the Antarctic Peninsula basalts differ from both in the above large-ion lithophile (LIL) element characteristics. These distinctions become especially significant in helping to define the effects of mantle plume activity in the West Antarctic rift system, as described in a later section.

Felsic rock types associated with the West Antarctic rift system include phonolite, trachyte, and several varieties of rhyolite, all of which are commonly peralkaline. However, the large felsic volcanoes tend to be either predominantly phonolitic (e.g., Mount Erebus and Mount Sidley), predominantly trachytic (Mount Melbourne and Mount Berlin), or predominantly rhyolitic (e.g., Mount Andrus). Isotopic and other geochemical studies all have concluded that the felsic and intermediate rocks evolved mainly by fractional crystallization of a basaltic parent and that crustal contamination and anatexis were of minimal significance [*Goldich et al.*, 1975; *Stuckless and Ericksen*, 1976; *Kyle*, 1981; *Wörner et al.*, 1989]. Evidently the conditions in this rift system were not suitable for the production of large-volume felsic ignimbrites like those found in the Rio Grande and East African rift systems.

In Marie Byrd Land, felsic volcanic activity has migrated centrifugally outward along rectilinear ranges from the center of the province toward its perimeter during the past 18 m.y., while basaltic activity has shown no chronologic pattern of spatial distribution. This has been interpreted as a manifestation of crustal doming along the flanks of the rift system, related in turn to the rise of an asthenospheric diapir or plume [*LeMasurier and Rex*, 1989].

In general terms, the picture that seems to be emerging for rift-related magmatic activity in this region involves (1) the generation of alkaline basaltic magma from mantle sources, largely within the asthenosphere; (2) eruption of a large proportion of this material without major modification; and (3) differentiation of other batches of basaltic magma in crustal reservoirs by a complex variety of fractional crystallization processes.

Chronology

Cenozoic volcanism associated with rifting seems to have begun roughly in late Eocene to middle Oligocene time in the Marie Byrd Land and Ross Embayment areas but did not begin until late Pliocene time at the Bellingshausen Sea end of the rift. Drill cores (Cenozoic Investigations of the Western Ross Sea (CIROS)) from McMurdo Sound record volcanic activity of Oligocene or older age [*George*, 1989], and the oldest exposed alkaline rocks along the Transantarctic Mountains front are 18–25 Ma [*Schmidt-Thomé et al.*, 1990]. In Marie Byrd Land, basaltic volcanism seems to have begun around 25–30 Ma [*LeMasurier*, 1990*a*, *b*] and continued with few interruptions; the oldest felsic volcano formed around 18–20 Ma [*LeMasurier*, 1990*b*]. In the Bellingshausen volcanic province [*Rowley et al.*, 1990] the oldest basalts are 7–10 Ma in Ellsworth Land and about 7 Ma [*LeMasurier*, 1990*a*, *b*] in northern Alexander Island. Although Mount Erebus (Ross Island) and Mount Melbourne [*Wörner and Viereck*, 1989] are the only demonstrably active volcanoes in the rift system, 15 others are suspected of Holocene activity [*LeMasurier*, 1990*a*]. The distribution of ages suggests, therefore, that the rift system has propagated gradually toward the Bellingshausen Sea–Antarctic Peninsula while remaining active at the Ross Sea end and that no part of the rift system is inactive. This progression, if true, is significant and important to consider in any general model which might account for the tectonic activity of the West Antarctic rift system.

We do not, of course, know the age of volcanism beneath the Ross Sea shelf, Ross Ice Shelf, and Byrd Subglacial Basin inferred from the geophysical data discussed in the magnetics section. These geophysically interpreted volcanic rocks are substantially more extensive in area than the rocks exposed in uplifted areas (Figure 2).

GEOPHYSICAL INVESTIGATIONS IN WEST ANTARCTICA

We review the geophysical studies of the West Antarctic rift system, beginning with aeromagnetic data because it provides the widest coverage of the area, then discuss the available gravity profiles (also over a wide area), and finally turn to the generally sparse oversnow seismic investigations. In the Ross Sea, marine seismic studies are extensive.

AEROMAGNETIC SURVEYS

We discuss first the results from widely spaced flight lines magnetic (Figures 5 and 6) and oversnow traverses throughout the region. We then concentrate on our recent BGR-USGS investigations in the Ross Sea–northern Victoria land area (Figures 7–10 and Plates 1 and 2). We refer to recent geophysical surveys of the southern Antarctic Peninsula–Ronne Ice Shelf area by the British Antarctic Survey (BAS), as appropriate.

High-amplitude linear magnetic anomalies are commonly produced by exposed and unexposed volcanic rocks that are found in rift zones. In Antarctica, prominent magnetic anomalies [*Behrendt and Wold*, 1963; *Behrendt*, 1964*a*, *b*; *Behrendt and Bentley*, 1968; *Robinson*, 1964*b*; *Pedersen et al.*, 1981; *Jankowski et al.*, 1983; *Bosum et al.*, 1989; *Behrendt et al.*, 1991] mark both Cenozoic volcanic rocks in Marie Byrd Land and the Transantarctic Mountains and Early Jurassic mafic magmatic rocks in the Transantarctic Mountains (e.g., Figure 5 and Plate 1). In the Ross Sea area, our recent aeromagnetic survey (Figures 7–9 and Plates 1 and 2) provides much finer resolution of the magnetic signatures of these somewhat contrasting magnetic rock types than the earlier widely spaced surveys. For example, the exposed Ferrar dolerite sills produce low-amplitude anomalies compared with the Jurassic Kirkpatrick basalts and late Cenozoic volcanoes (Plate 1); therefore, we would not expect significant amplitude anomalies from sills buried beneath the ice sheet or continental shelf.

Aeromagnetic survey flights in the late 1950s and early 1960s [*Behrendt and Wold*, 1963; *Behrendt*, 1964*a*], which are still the only data available in much of West Antarctica, were made with poor navigational accuracy that produced horizontal errors distant from control points as great as 50 km. Although the actual magnetic data were measured to 1 or 2 nT, the widely spaced lines and lack of control for secular variation away from base stations made it impossible to produce reasonable contour maps of total intensity. The profiles were interpreted using mostly two-dimensional methods [e.g., *Behrendt and Wold*, 1963]. An aeromagnetic survey of Ross Island near McMurdo made by E. Thiel in 1958 with a 100-nT contour interval [*Robinson*, 1964*b*] showed prominent anomalies (e.g., 2000 nT over active Mount Erebus) produced by the late Cenozoic volcanic rocks that

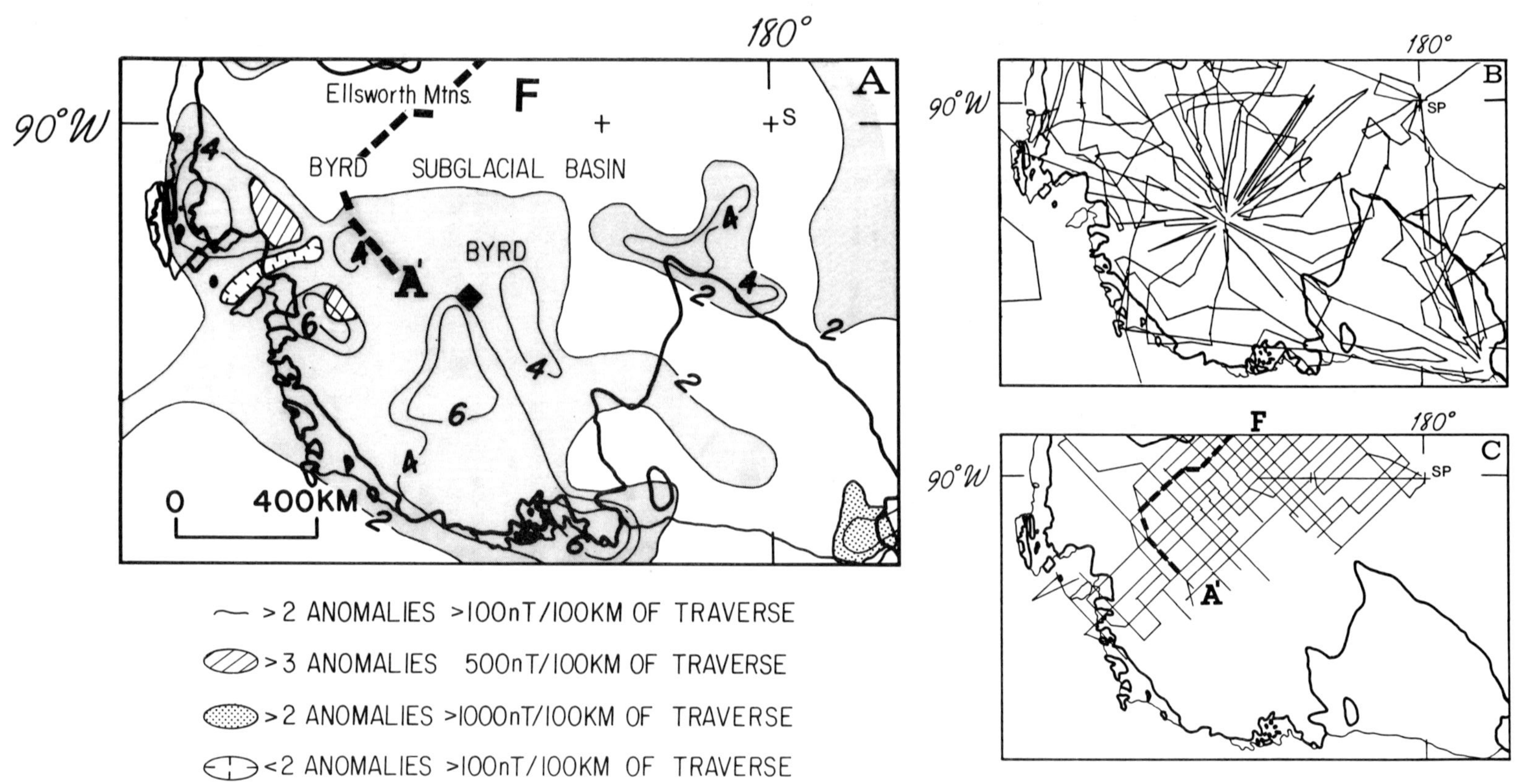

Fig. 5. (*a*) Frequency of occurrence of short-wavelength, high-amplitude anomalies. Contours show the number of anomalies greater than 100 nT per 100 km of flight line averaged for 1° of latitude grid (111 × 111 km). Modified from *Behrendt* [1968], using additional data collected in 1978 [*Jankowski et al.*, 1983; *Drewry*, 1983]. The location of profile A′—F (Figure 6) is indicated. The magnetic "break" is defined by ">2" contour in the Byrd Subglacial Basin. (*b*) Locations of flight lines used by *Behrendt* [1968]. (*c*) Location of flight lines collected in 1978 at approximately 100-km spacing, combined with radar ice sounding [*Jankowski et al.*, 1983; *Drewry*, 1983]. Location of profile A′—F (Figure 6 and Plate 2) is indicated by a heavy line. Grid north is at the top of Figures 5*a*–5*c*.

comprise Ross Island. *Pedersen et al.* [1981] repeated this survey and published a 200-nT contour map. Although the anomalies were published along profiles [*Behrendt and Wold,* 1963; *Ostenso and Thiel,* 1964; *Behrendt,* 1964*b*], a statistical method [*Behrendt,* 1964*a*; *Behrendt and Bentley,* 1968] allowed the display of general magnetic character over Antarctica, where profiles were variably spaced.

In 1978, inertial navigation (with an accuracy to about 5 km [*Jankowski,* 1981]) was used for aeromagnetic surveys combined with radar ice sounding [*Behrendt et al.,* 1980, 1981; *Jankowski et al.,* 1983; *Drewry,* 1983] (Figures 4 and 5) over part of the West Antarctic rift system and the Middle Jurassic Dufek intrusion. The flight lines were too widely spaced and were not controlled for time variation of the magnetic field in order to contour the data, so the anomalies were shown along profiles. In this paper, we combine these data as published by *Drewry* [1983] with the earlier compilation [*Behrendt* 1964*a*, 1968] to produce the map shown in Figure 5. We did not have available the recently published BAS [*Maslanyj and Storey,* 1990] data, but in the area of Figure 5 the results would not be changed had they been included.

Evidence for Cenozoic Volcanic Rocks Beneath Ice

The obvious correlation of high-amplitude short-wavelength magnetic anomalies with outcrops of late Cenozoic volcanic rocks in the Marie Byrd Land area (Amundsen-Bellingshausen flank) (Figures 2 and 5) was noted earlier [*Behrendt and Wold,* 1963; *Behrendt,* 1964*a*]. Numerous magnetic anomalies, which we infer are caused by subglacial late Cenozoic volcanic rocks, extend over a broad area of Marie Byrd Land and the Byrd Subglacial Basin far from volcanic outcrops. The prominent (>1000 nT) linear magnetic "Sinuous Ridge" was interpreted as caused by volcanic rocks of unknown age [*Jankowski et al.,* 1983] (Figure 2). Possibly the Sinuous Ridge anomaly in the deepest part of the Byrd Subglacial Basin represents a change from alkaline to tholeiitic magmatism and the start of formation of an oceanic spreading center in the late Cretaceous or Cenozoic, although there is no gravity or other supporting geophysical evidence for this. We interpret most anomalies in the area covered within the closed "2" contour of Figure 5 over the ice-covered area of Marie Byrd Land (Byrd Subglacial Basin) where depths to sources are <1 km below the base of the ice [*Behrendt and Wold,* 1963; *Jankowski et al.,* 1983] to be the result of late Cenozoic volcanism similar to that of the volcanic ranges of Marie Byrd Land (Figure 2), although some are certainly caused by older shallow sources. Locations of these ranges are shown by *LeMasurier and Rex* [1989]. Specifically Mount Takahe, Kohler Mountains, Hudson Mountains, Crary Mountains, Mount Toney, and Mount Sidley have associated high (100–500 nT) short-wavelength anomalies correlated with ridges and peaks [*Behrendt and Wold,* 1963; *Ostenso and Thiel,* 1964]. In contrast, the basement granitic and metamorphic rocks exposed in Marie Byrd Land have few or no magnetic anomalies where crossed by the profiles shown in Figure 5 [*Behrendt,* 1964*a*; *Ostenso and Thiel,* 1964], except for the anomalies over metamorphic basement within the "6 anomalies >100 nT per 100 km of traverse" contour at the bottom of Figure 5*a*.

Magnetic Boundary in Byrd Subglacial Basin

The character of the magnetic field changes abruptly, as marked by the north trending part of the ">2" contour in Figure 5*a*, about half way between the outcropping volcanoes of Marie Byrd Land and the Ellsworth Mountains (Figures 5 and 6, a representative profile). This "break" was first noted on an oversnow traverse by *Bentley and Ostenso* [1961] and described by *Behrendt and Wold* [1963] and is quite

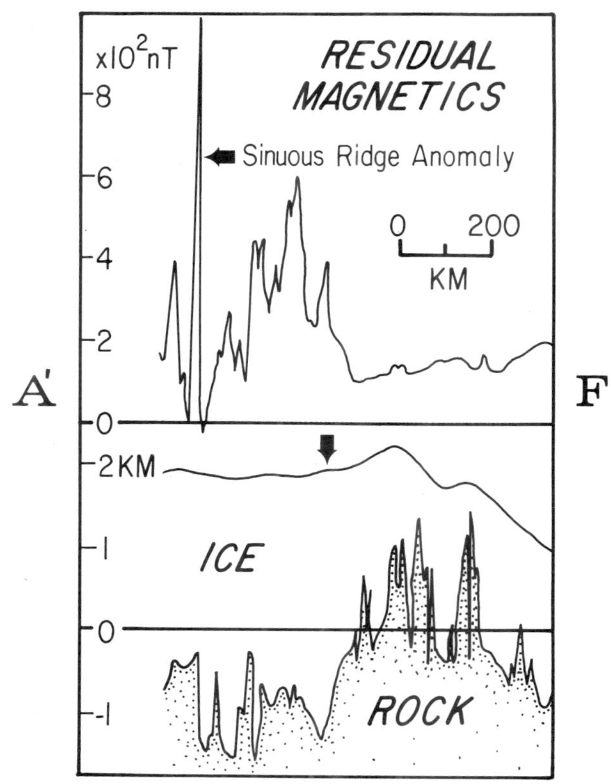

Fig. 6. Radar ice-sounding and aeromagnetic profile across the extended crust of Byrd Subglacial Basin and near the Ellsworth Mountains, from *Drewry* [1983]. Vertical exaggeration of the topography is 200:1. The arrows mark the interpreted rift shoulder and also the "break" in magnetic anomalies. The magnetic break (also Figure 5) separates a region of numerous high-amplitude anomalies, interpreted as caused mostly by upper Cenozoic volcanic rocks as sources (90% ≤ 1 km below bedrock surface), contrasted with nonmagnetic rock (possibly granitic but most likely metasedimentary because of exposures in the Ellsworth Mountains). The ~1000-nT anomaly is the Sinuous Ridge anomaly (Figures 2 and 5). The location of the profile is indicated in Figure 5.

apparent in other profiles [e.g., *Drewry*, 1983]. Although some of the shallow sources of these anomalies may be pre-Cenozoic mafic magnetic rock close to the base of the ice as well as Cenozoic volcanic rocks, the absence of shallow magnetic sources on the Ellsworth Mountains side of the "break" depicted in Figure 5 is good evidence for the lack of Cenozoic volcanic rocks beneath the ice sheet there.

Interpretation of magnetic depths along the magnetic-radar-ice–sounding profiles in Figure 5*c* combined with the earlier determined depths [*Behrendt and Wold,* 1963] were used to compile maps [*Jankowski et al.*, 1983] showing depth to magnetic basement and overlying thickness of nonmagnetic rock. On these maps (not shown), most depth to magnetic basement is less than 1 km in the area of steep short-wavelength anomalies within the "2" contour of Figure 5, including the Byrd Subglacial Basin. Seismic refraction data, discussed later, support the magnetic depth and volcanic source interpretation. The area toward the Ellsworth Mountains from the magnetic break in Marie Byrd Land mostly contains nonmagnetic rock as thick as 5 km, with individual depth determinations as great as 7–11 km [*Jankowski et al.*, 1983].

A profile across part of the West Antarctic rift and rift shoulder (Figure 6) illustrates the magnetic break between the region of shallow, strongly magnetic rocks (magnetic basement close to the bedrock surface) interpreted as caused by late Cenozoic volcanic rock and the region underlain by thick nonmagnetic rocks which probably extends into the Ellsworth Mountains. The rugged subglacial topography is also continuous with that of the precipitous Ellsworth Mountains. Figure 6 illustrates the fact that the magnetic break (Figure 5) which we infer to mark the limit of late Cenozoic volcanic rocks in this area may also approximately mark the inferred edge of the rift shoulder (arrow). However, in the Ross Sea area and the Amundsen-Bellingshausen flanks of the West Antarctic rift system the magnetic Cenozoic rocks extend into the rift shoulders, so we cannot interpret the magnetic pattern to mark the rift shoulder where buried.

In contrast to the aeromagnetic anomalies over Marie Byrd Land (including the Byrd Subglacial Basin), which we correlate with late Cenozoic volcanic rocks, other high-amplitude, short-wavelength anomalies elsewhere over ice-covered areas are more problematic to interpret. *Behrendt* [1964*a, b*], *Behrendt et al.* [1974, 1980, 1981, 1990], *Garrett et al.* [1987, 1988], *Dalziel et al.* [1987], and *Bosum et al.* [1989] discuss magnetic anomalies over the Transantarctic Mountains (including the Dufek intrusion), Whitmore Mountains area, and southern Antarctic Peninsula. The interpreted intrusive and extrusive magnetic rocks inferred to cause the anomalies range from Precambrian [*Garrett et al.*, 1987; *Maslanyj and Storey,* 1990] to Middle Jurassic [e.g., *Behrendt et al.*, 1981] in both the southern Antarctic Peninsula and Transantarctic Mountains. We cannot differentiate between pre-Cenozoic igneous rocks and late Cenozoic volcanic rocks on the basis of magnetic data alone. We point out, however, that there are late Cenozoic volcanic outcrops in the southern Antarctic Peninsula (Figure 2) and note the position of several linear anomalies each (about 1000 nT in intensity) indicated in Figure 2, similar to the Sinuous Ridge anomaly [*Jankowski et al.*, 1983] in the Byrd Subglacial Basin, which we suggest is caused by the rift-related Cenozoic(?) volcanism.

Anomalies in Haag Nunataks–Southern Antarctic Peninsula Area

Garrett et al. [1987] showed three linear anomalies (H in Figure 2, which designates location near Haag Nunataks) which are subparallel to and south of a similar 1300-nT anomaly (A in Figure 2) [*Behrendt,* 1964*b*]. *Garrett et al.* [1987] suggested that the high uplift of the northern Ellsworth Mountains (Sentinel Range) was a Cenozoic horst (which fits the interpretations of this paper) but interpret the sources of the linear anomalies (H in Figure 2) between the Ellsworth Mountains and the southern Antarctic Peninsula as being Precambrian in age because magnetic rock of that age is exposed at Mount Haag. They suggest Cenozoic rifting possibly separated the southern Antarctic Peninsula from the Ellsworth Mountains, thus extending the interpretation of this topographic tectonic break [*Behrendt,* 1964*b*; *Doake et al.*, 1983]. Considering that late Cenozoic volcanic rock [*Rowley et al.*, 1990] crops out about 50–100 km northeast of Mount Haag in the Merrick Mountains (not labeled) (75°06′S, 72°04′W), we suggest that Cenozoic mafic (or ultramafic?) magmatic rocks are possible sources for some of these high-amplitude linear anomalies (A, S, and H in Figure 2). *Maslanyj and Storey* [1990] reported recent BAS survey data over the Ellsworth–Whitmore Mountains, southern Antarctic Peninsula, and Weddell Embayment, mostly out of the area of this paper. Their results, however, did improve the definition of the Haag Nunataks and Antarctic Peninsula traverse anomalies, indicating greater complexity than the simple linear trends we indicate at the scale of Figure 2. Their data over the Ellsworth Mountain area are generally consistent with the smooth field indicated in Figure 5.

ROSS SEA SHELF–NORTHERN VICTORIA LAND AEROMAGNETIC SURVEY

Figure 7*a* shows an index map of this well-studied area. Widely spaced magnetic profiles over the Ross Sea shelf and adjacent part of the Transantarctic Mountains [*Behrendt,* 1964*a*, 1968; *Behrendt et al.*, 1987] show numerous high-amplitude (300–2000 nT), short-wavelength anomalies over marine and ice-covered areas. Although Jurassic bodies several kilometers thick such as the Dufek intrusion [*Behrendt et al.*, 1974, 1981] or the Kirkpatrick basalts in the

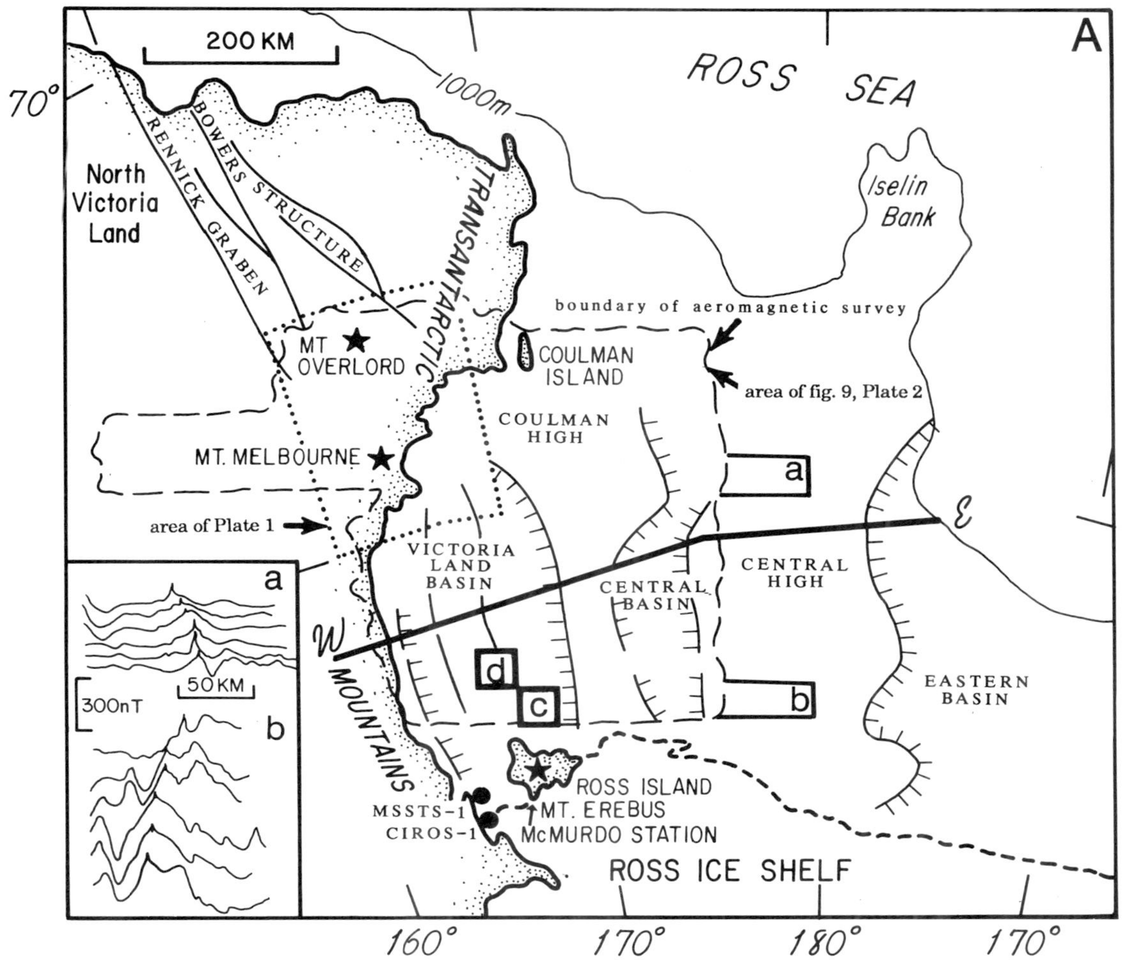

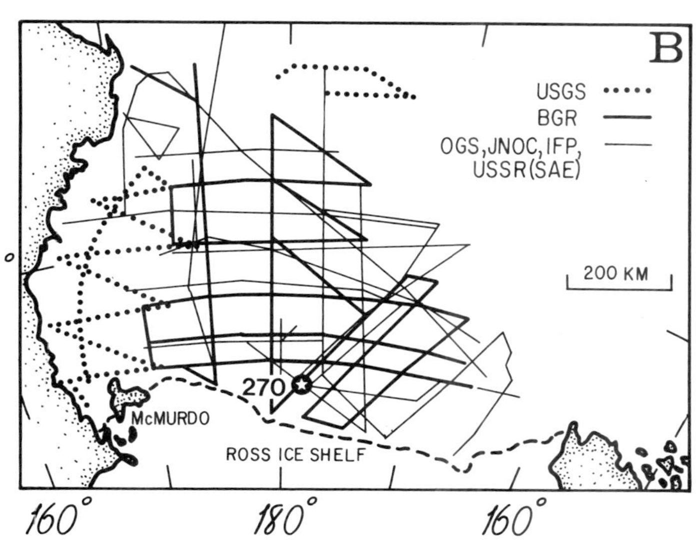

Fig. 7. (*a*) Index map of structures and geographic locations in the Ross Sea shelf–northern Victoria Land area. Active volcanoes Mount Erebus and Mount Melbourne and late Cenozoic volcano Mount Overlord (Pleistocene?) are indicated. Boxes a and b indicate the locations of the stacked magnetic profiles a and b shown in the lower left of this figure. (Ticks indicate anomaly correlations between profiles.) (Structures are from *Davey et al.* [1982, 1983] and *Cooper et al.* [1987*a*].) Locations of profiles C and D in Figure 10 and the MSSTS-1 and CIROS-1 core holes and the approximate location of the cartoon in Figure 17, profile W—E, are indicated. A 1000-m bathymetric contour is shown. (*b*) Locations of seismic reflection profiles over the Ross Sea shelf. The profiles from the U.S. Geological Survey and Bundesanstalt für Geowissenschaften und Rohstoffe (USGS and BGR) (also indicated in Plate 2) are used in interpretations in this paper. Other profiles include the Institut Français du Pétrole (IFP), France; Japanese National Oil Company (JNOC), Japan [*Sato et al.*, 1984]; Soviet Antarctic Expedition (SAE), USSR; and Oservetoreo Geofisico Sperimentale (OGS), Italy. Deep Sea Drilling Project core hole 270 is also indicated. Maps in Figures 7, 8, 9, and 15 and Plates 1 and 2 (all covering the Ross Sea–northern Victoria Land area) have normal map conventions for true north, approximately toward the top of each map.

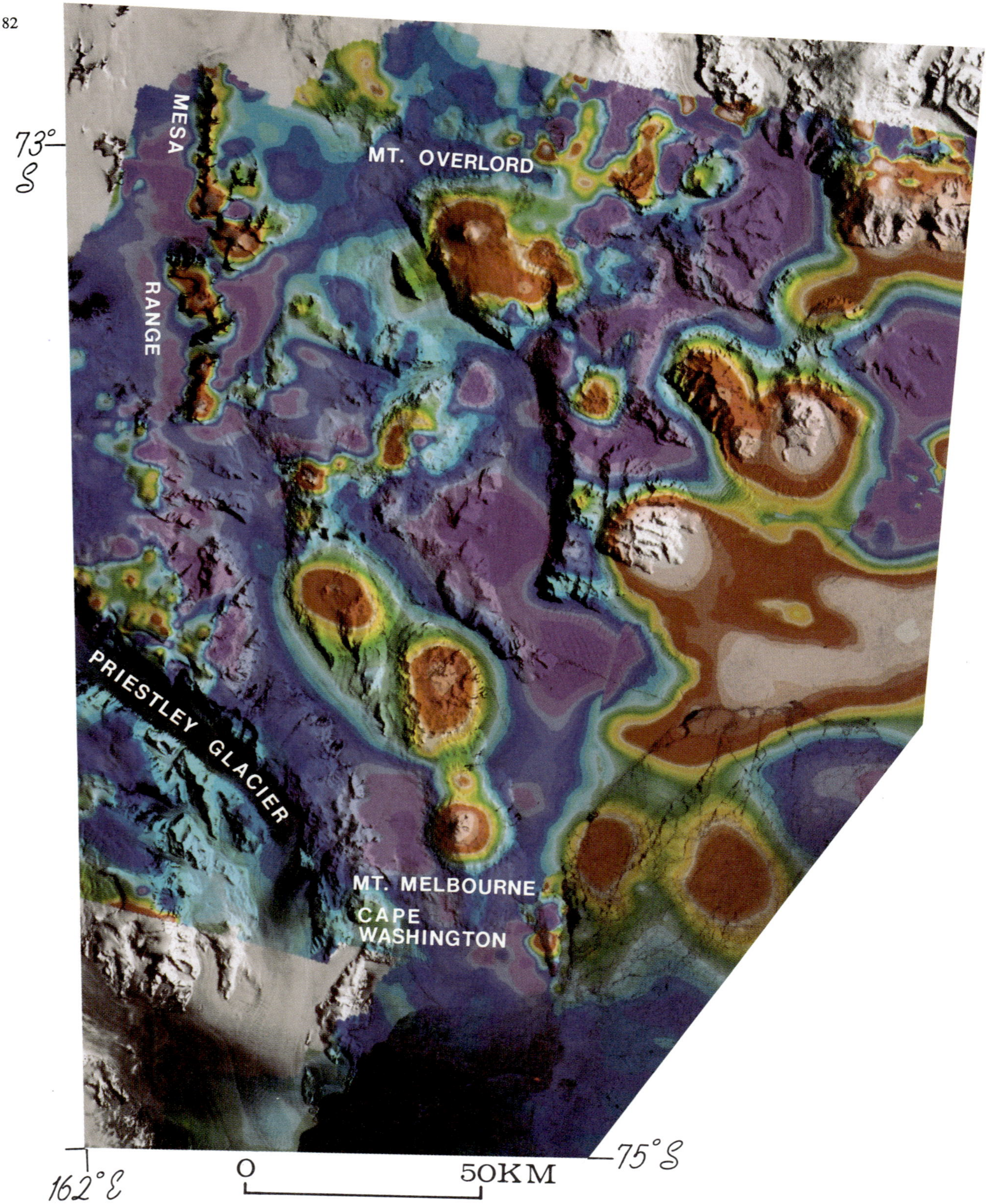

Plate 1. Aeromagnetic survey over part of northern Victoria Land, superimposed on enhanced Landsat band 7 (compiled by B. Lucchitta, U.S. Geological Survey). The area is shown in Figure 7. The contour interval is 5 nT except for the highest amplitude, as in Plate 2. Active volcano Mount Melbourne and late Cenozoic volcano Mount Overlord are indicated. Exposures of Jurassic Kirkpatrick basalts in the Mesa Range have several hundred nanotesla associated anomalies. At the north edge of Priestley Glacier, Ferrar dolerites are exposed in cliff faces [*Roland et al.*, 1989], but only a few tens of nanotesla anomalies are produced by the flat-lying sills. The small circular high-amplitude anomalies just north of the edge of Priestley Glacier are probably caused by small exposures of Kirkpatrick basalt (N. Roland, personal communication, 1989). Cape Washington consists of late Cenozoic volcanic rocks [*Wörner and Viereck*, 1989] and has an associated magnetic anomaly. The two approximately 10-km-wide circular anomalies 20 to 60 km northwest of Mount Melbourne must be caused by sources (intrusions?) buried beneath the nonmagnetic exposed rock [*Bosum et al.*, 1989].

Mesa Range (Plate 1) produce high-amplitude anomalies, we speculate that the high-amplitude, shallow-source, short-wavelength anomalies over the Ross Sea shelf (Figure 5) and grounded ice at its inland end are predominantly caused by late Cenozoic volcanic rocks (having substantial vertical thickness) based on comparison of these anomalies to similar anomalies over young volcanic exposures in other areas [*Behrendt*, 1964*a*; *Robinson*, 1964*a*; *Pederson et al.*, 1981]. The possibility that these magnetic anomalies might be caused by Jurassic Ferrar–type dolerite sills is rejected because, although these bodies are quite magnetic, the sills are essentially thin sheets, produce only small anomalies and only at their edges, and do not produce anomalies greater than about 50–100 nT at the flight elevations (Plate 1) (also, for discussion of models for sills, see *Pederson et al.* [1981]). Therefore it is understandable that the numerous profiles crossing the Transantarctic Mountains (Figure 5) directly over dolerite sills [*Behrendt*, 1964*a*] did not record significant high-amplitude anomalies.

A 4.4 km × 20 km spaced aeromagnetic survey was flown over northern Victoria Land and the southwestern Ross Sea [*Bosum et al.*, 1989; *Behrendt et al.*, 1991]. These data [*BGR-USGS*, 1987], contoured at a 5-nT interval, provide substantially more information

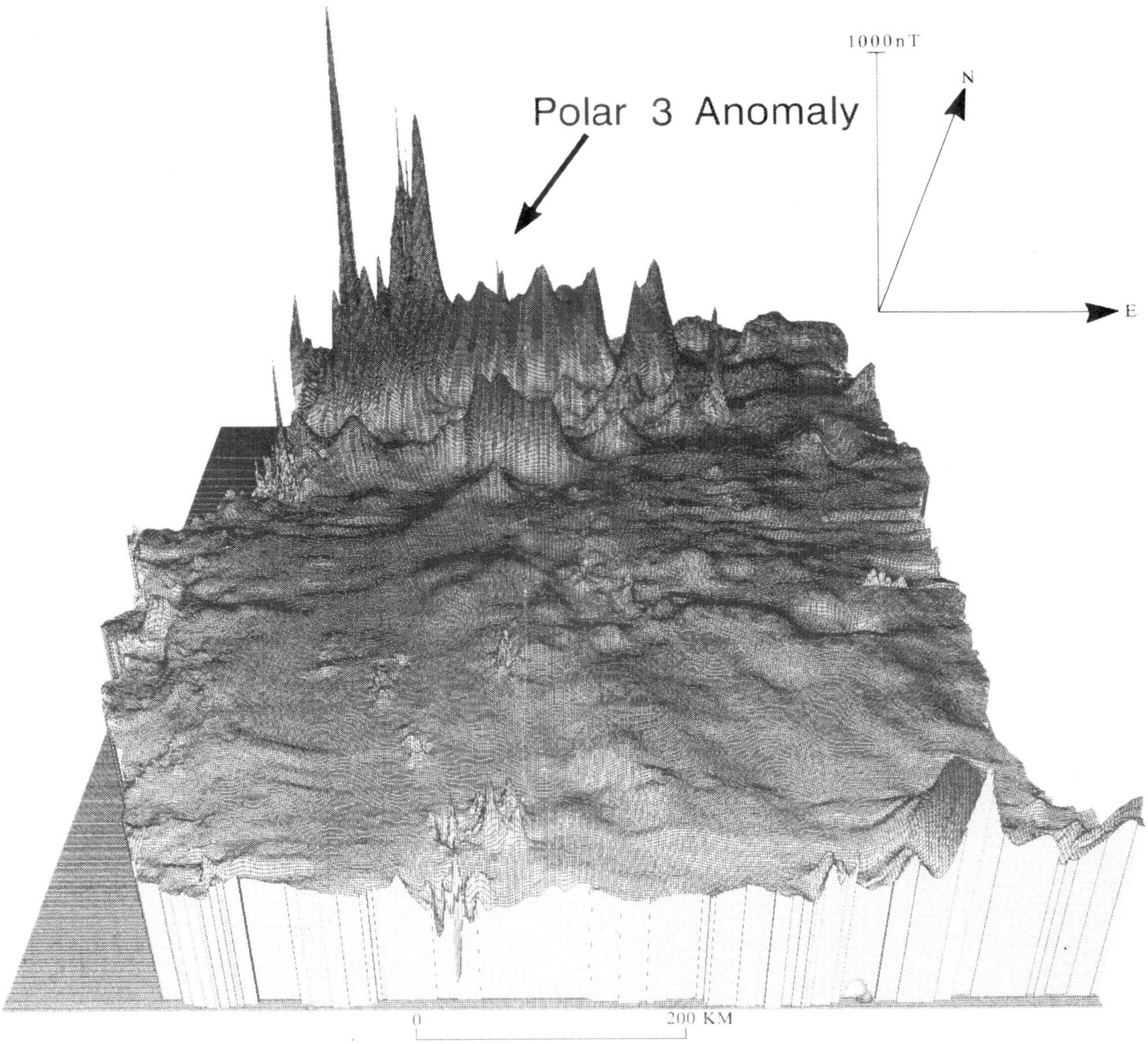

Fig. 8. A three-dimensional perspective of the total magnetic intensity map of the marine area east of the coast depicted in Plate 2, viewed from the south. Amplitudes are projected at the vertical scale shown. Geologically significant anomalies indicated here and in Plate 2 range over 3 orders of magnitude in amplitude and wavelength (compiled by D. Damaske).

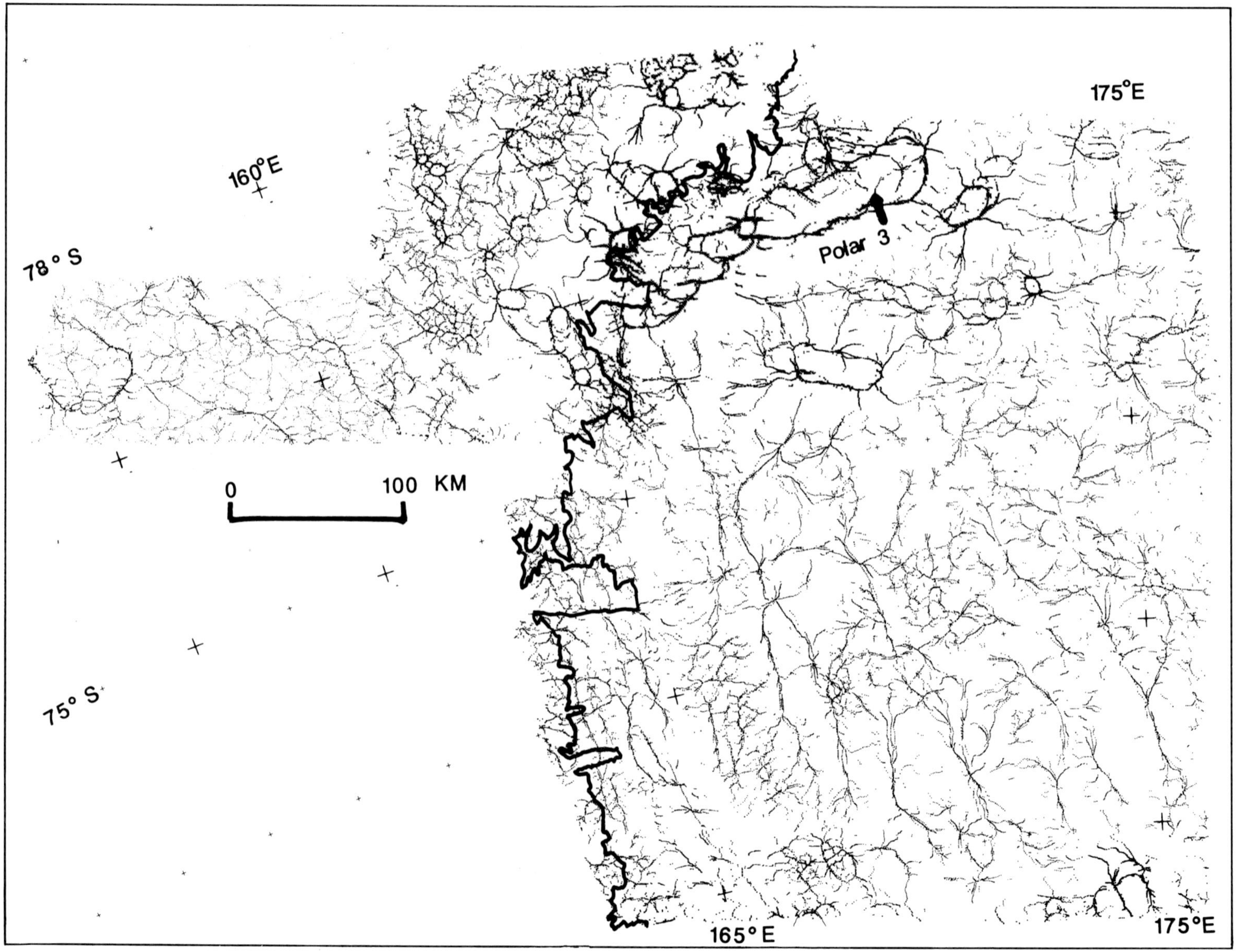

Fig. 9. Local maxima of the horizontal gradient of pseudogravity calculated from the aeromagnetic data found in Plate 2 (for explanations of methods, see *Cordell and Grauch* [1985] and *Blakely and Simpson* [1986]). This calculation procedure (which has nothing to do with gravity) normalizes the anomaly amplitudes and areally defines the edges of the bodies producing the magnetic anomalies. The lines drawn on this figure are the loci of inflection points and mark the edges of bodies (compiled by R. Saltus).

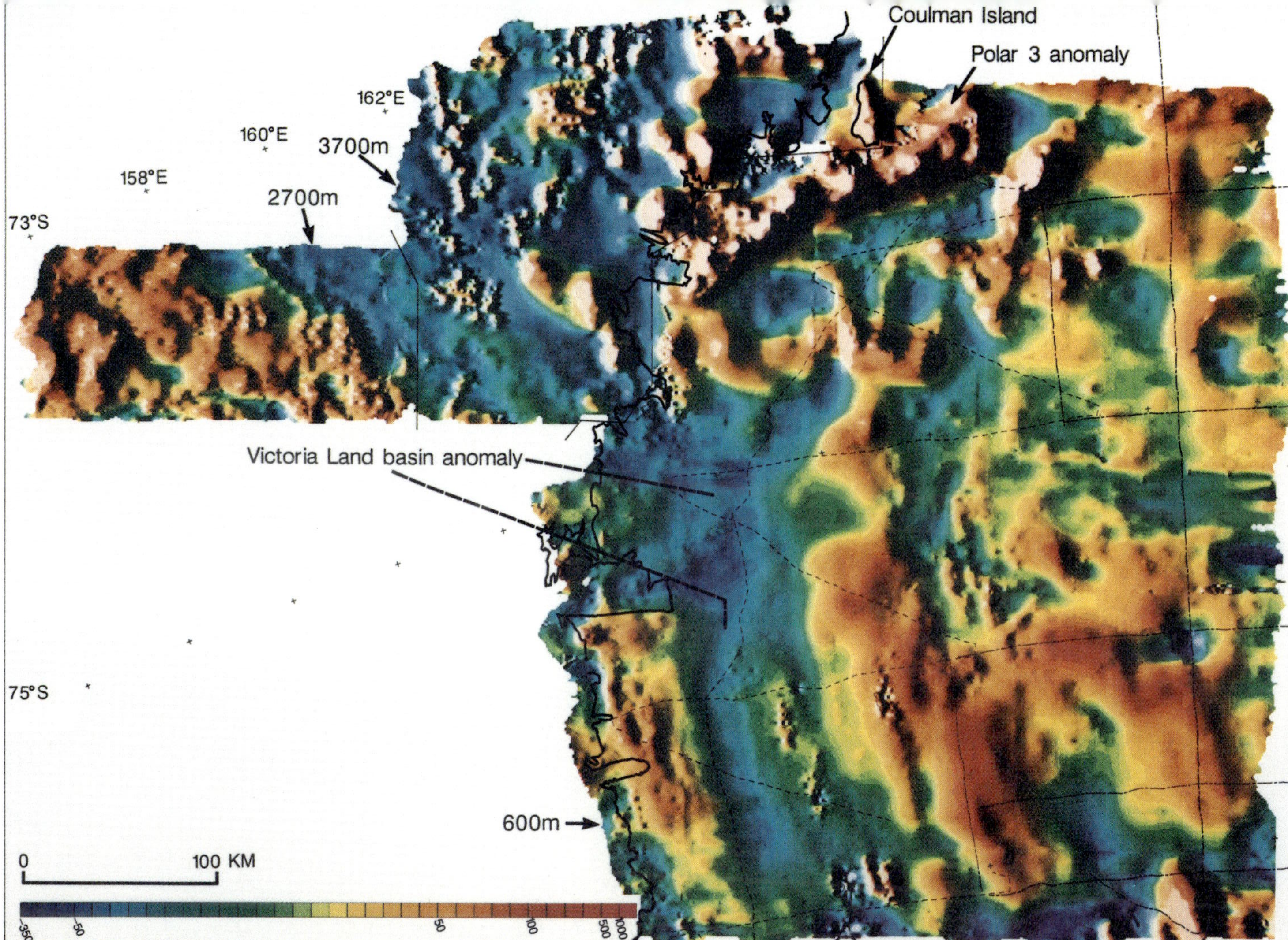

Plate 2. Color-shaded relief (apparent sun angle 15° from west) and aeromagnetic map of western Ross Sea shelf (right half of figure) and part of northern Victoria Land (top central and left part of figure). Location of the survey is shown in Figure 7. Areas flown at different elevations are separated by heavy solid-line segments: 600 m, Ross Sea; 3700 m, northern Victoria Land; 2700 m, East Antarctic ice sheet at west (far left). Flight lines were spaced 4.4 km (EW) (some at 8.8 km) by 22 km (NS). This map was compiled using an 800 × 800 m grid. Locations of marine seismic reflection profiles collected by the U.S. Geological Survey's R/V *S.P. Lee* in 1984 (short dash lines) and the *Explora* of the Bundesanstalt für Geowissenschaften und Rorstoffe in 1980 (dash-dot lines) are indicated. The north tip of Ross Island is present at the south edge of the map; the coastline is indicated by the solid line. The contour interval is 5 nT between +50 and −50 nT and compressed, as indicated in color bar beyond this range. The Polar 3 anomaly was probably caused by upper Cenozoic volcanic rock and a subvolcanic (mafic?) intrusion. The −80 to −100 nT north trending anomaly 80 to 100 km wide at the southwest corner of the map overlies the Victoria Land basin. The elongate western area of magnetic coverage (2700 m) overlies the East Antarctic ice sheet. The major northwest trending break in magnetic character at the right of this area probably marks the east edge of the buried Precambrian shield.

per unit area than that known for most of the West Antarctic rift system. Geologically significant individual anomalies defined by this survey range in amplitude from 10–20 nT up to 1700 nT and in length from 1–2 km to about 200 km. Figures 7–9 and Plates 1 and 2 illustrate these features, which include (1) the numerous small circular anomalies interpreted as submarine, late Cenozoic volcanic edifices and subvolcanic intrusions; (2) the 1700-nT, 200-km-long Polar 3 anomaly (centered about 74°S, 168°E); and (3) the −80- to −100-nT anomaly over the Victoria Land basin.

Figures 8 and 9 and Plate 2 show different presentations of the data set from this survey and thus should be examined together. For example, the shaded relief (Plate 2) accentuates rift fabric, but short-wavelength anomalies are hard to see at this small scale. The 5-nT contour interval present over most of the map defines well the longer-wavelength low-amplitude anomalies; however, anomalies greater than 100 and 500 nT tend to become obscured in the red shades. In contrast, Figure 8 provides the best subjective illustration of amplitude and gradient (which suggests depth to source) for all the anomalies in the Ross Sea area. However, the locations of anomalies are distorted by the three-dimensional presentation. Figure 9 provides the most information on location of edges of anomaly-producing bodies irrespective of depth. This type of presentation is somewhat analogous to a seismic epicenter map in which all data are projected to the surface irrespective of depth. In general, shapes of short-wavelength circular anomalies are not well resolved in Plate 2 and Figure 9 because of the limitations resulting from wide (4.4 km) flight lines compared with the smallest anomaly width of 1 to 2 km, which would only be crossed by one flight line, even considering the 700-m flight elevation and the 500–1400 m deep continental shelf.

Anomalies Caused by Late Cenozoic Volcanoes

The steeper gradients of circular anomalies caused by volcanoes like Mount Overlord and Mount Melbourne (Plates 1 and 2) can be compared with those caused by similar but lower-gradient circular sources within the lower Paleozoic basement that can be seen both onshore and offshore (Plate 1). Cenozoic intrusions within competent crystalline basement rock probably could produce these circular anomalies (Plates 1 and 2) which have diameters of >10 km. Similar volumes of the same magma penetrating the thick sedimentary rock within the Victoria Land basin would produce sills with low-amplitude anomalies and the smaller-diameter (1 to 2 km) circular anomalies such as those observed over vertical bodies which underlie submarine volcanoes (e.g., Figure 10). However, older intrusions could produce sources of these ~10-km-diameter circular anomalies, some of which are not exposed (Plate 1), over outcropping basement.

Ten small anomalies having short (1–5 km) wavelengths and high (a few tens to more than 1000 nT) amplitudes were found along seismic reflection profiles using magnetic gradiometer data (e.g., Figure 10). They are attributed to late Cenozoic subvolcanic intrusions [*Behrendt et al.*, 1987] that penetrate a thick (up to 14 km) sedimentary section. The aeromagnetic survey defines many more (about 100) similar type small anomalies (Figure 8 and Plate 2) which may also have an analogous cause. Based on theoretical magnetic models compared with seismic reflection profiles for several of the very short wavelength (1–3 km) anomalies along ship tracks, we infer that causative bodies for many of these 100 small anomalies penetrate from the basement through the sedimentary section. Depth estimates indicate that the tops of the sources are essentially at the seafloor and probably would have topographic expression like those in Figure 10 were seismic or bathymetric profiles measured over them. However, only in areas of the Ross Sea floor underlain by late Cenozoic sedimentary rock [*Behrendt et al.*, 1987] can we infer that the ages of the (shallow) magmatic rock are late Cenozoic or Holocene (Figures 10*c* and 10*d*, respectively). These penetrating bodies have been defined by available seismic reflection data only in the Victoria Land basin area.

Magnetic Evidence of Rift Fabric Beneath Ross Sea Shelf

The small magnetic anomalies do not occur randomly over the marine area but are concentrated along linear zones mostly parallel to the Victoria Land basin (Figure 7 and Plate 2). We interpret this lineation as evidence of rift fabric. Note the north-northwest "grain" over the Ross Sea part of the survey, which can be better defined by locating the edges of anomaly-producing structures (Figure 9). Probably many of the north-northwest trending magnetic lineations are caused by faults in the magnetic basement, some of which are discernible on reflection profiles [*Cooper et al.*, 1987*a*]. This fabric can be seen across the survey area over the Victoria Land basin, Coulman high, Central basin (Figure 9 and Plate 2), and Central high (Figure 7). The short-wavelength (1 to 2 km) anomalies discussed above occur along some of these linear trends. This linear fabric (Figure 9 and Plate 2) could be, however, either Cenozoic or late Mesozoic in age, based on interpretation [*Cooper et al.*, 1987*a*] of several episodes of rifting.

The zone of lineations of north-northwest trending anomalies observed in Plate 2 can be extended nearly to the edge of the Eastern basin (Figure 7*a*) where additional east trending profiles are shown. We interpret the 1–3 km wavelength anomalies observed on the stacked profiles as shallow source (essentially at the seafloor) features, similar to anomalies to the west discussed above, on the basis of the depth

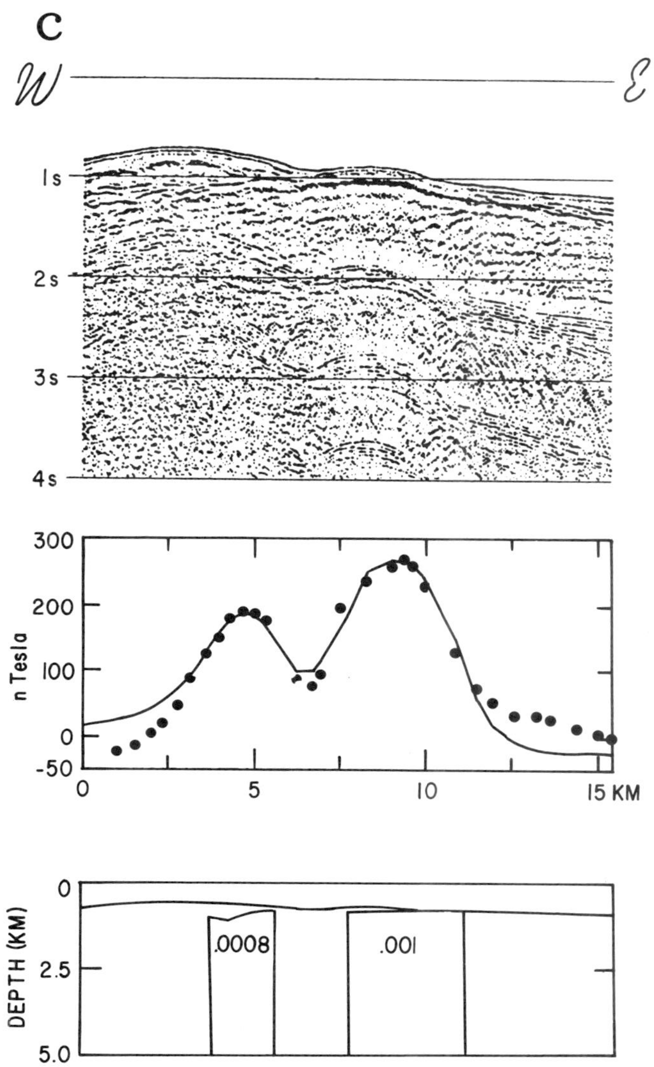

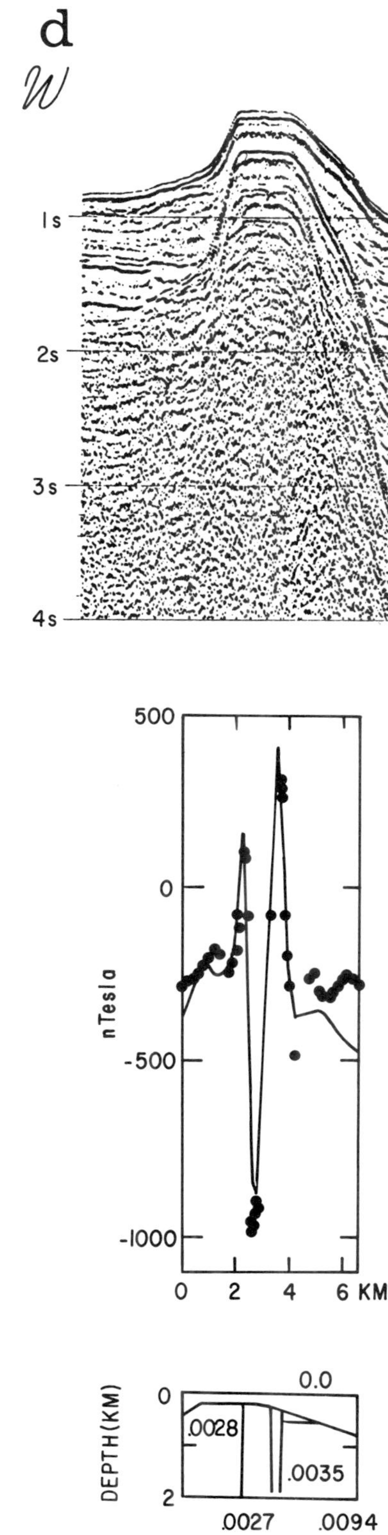

Fig. 10. (Left) Anomaly c [*Behrendt et al.*, 1987]. Seismic reflection profiles (time in seconds indicated) compared with observed (dots) and theoretical (solid line) marine magnetic profiles from *Behrendt et al.* [1987]. Apparent susceptibilities of theoretical models are shown in cgs units. Sources are interpreted to be upper Cenozoic subvolcanic intrusions. The smooth submarine topography across anomaly c probably indicates that former volcanic edifices were removed by glacial erosion. Therefore the intrusion is younger than the Neogene sedimentary rock but older than the latest deglaciation. (Right) Anomaly d [*Behrendt et al.*, 1987]. The prominent submarine volcano is probably Holocene; i.e., the edifice was extruded on the seafloor since deglaciation. The locations of anomalies c and d are shown in Figure 7.

estimates (horizontal extent of maximum gradient); but because these profiles cross the Central high only, we cannot say much about their age. A broad (about 40 km wide) north-northeast trending anomaly indicated by tick marks in box b of Figure 7*a* and having several sharp peaks exceeds 300 nT and indicates a shallow source.

Polar 3 Anomaly

The Polar 3 anomaly has sharp peaks (Plate 2 and Figure 8) similar to those over late Cenozoic volcanic rocks exposed on Coulman Island (Figure 7 and Plate 2), and thus we interpret the Polar 3 anomaly as a magmatic complex. Although generally linear, the curved edges of individual circular volcanic-intrusive centers can also be seen in Figure 9. *Bosum et al.* [1989] interpret serpentinized ultramafic rocks as one possible source. Probably the source consists of both volcanic deposits and subvolcanic intrusions. Transfer faults have been suggested to occur along the Transantarctic Mountains [*Fitzgerald*, 1989] and possibly the Polar 3 anomaly is related to these faults tectonically [*Wörner et al.*, 1989; *Cooper et al.*, 1991] because it has a similar trend. The Polar 3 anomaly is similar in amplitude and wavelength to other anomalies indicated in Figure 2 such as the Sinuous Ridge (Figures 2, 5, and 6) [*Jankowski et al.*, 1983], adjacent to the subglacial trench in the ice-covered areas of the West Antarctic rift system in Marie Byrd Land.

Other shallow-source magnetic anomalies occur over the Coulman high between the Victoria Land basin and Central basin (about 75°45′S, 170°E) (Figure 7 and Plate 2); these may be due to volcanoes and associated subvolcanic intrusions. Seismic velocities as high as 6.9–7.4 km/s are reported [*Cooper et al.*, 1987*b*] at shallow depths in this area as discussed below, which implies high-density mafic intrusions. We suggest the source of the 500-nT anomaly at the southeast corner of Plate 2 and Figure 8 is late Cenozoic volcanic rock.

Victoria Land Basin Anomaly

The source of the −80- to −100-nT anomaly over the Victoria Land basin, inferred to be filled with a 14-km-thick sedimentary sequence [*Cooper et al.*, 1987*a*], has been interpreted in two ways. One model fit to a profile across this anomaly suggests a deep (about 12–14 km) magnetic basement overlain by nonmagnetic sedimentary rock [*Behrendt et al.*, 1987, 1991]. A second model fit to the same profile is interpreted to represent a shallow Curie isotherm resulting from late Cenozoic rifting and a thinner (6–8 km) sedimentary section overlying rock possibly consisting of thermally demagnetized volcanic flows [*Behrendt et al.*, 1991]. High heat flow in the Victoria Land basin area is suggested by sparse and inconclusive measurements [*Blackman et al.*, 1987], active volcanism, tectonic activity (Figure 11), thermobarometry [*Berg and Herz*, 1986], a 40°C/km thermal gradient [*White*, 1989], and high seismic velocity at shallow depth within the reflector sequence interpreted as evidence of volcanic layers (see below). These features lend support to the shallow Curie isotherm interpretation.

GRAVITY SURVEYS

Although gravity data were routinely collected on the (1956–1966) oversnow traverse program (1955–1964) throughout the West Antarctic rift system, seismic reflection determinations of ice thickness measurements at only every eighth or tenth station (about 30–40 km) and poor elevation control (absolute errors about ±50 m, corresponding to ±15 mGal error in free-air anomaly [*Bentley*, 1964]) limit their usefulness beyond ice thickness determination

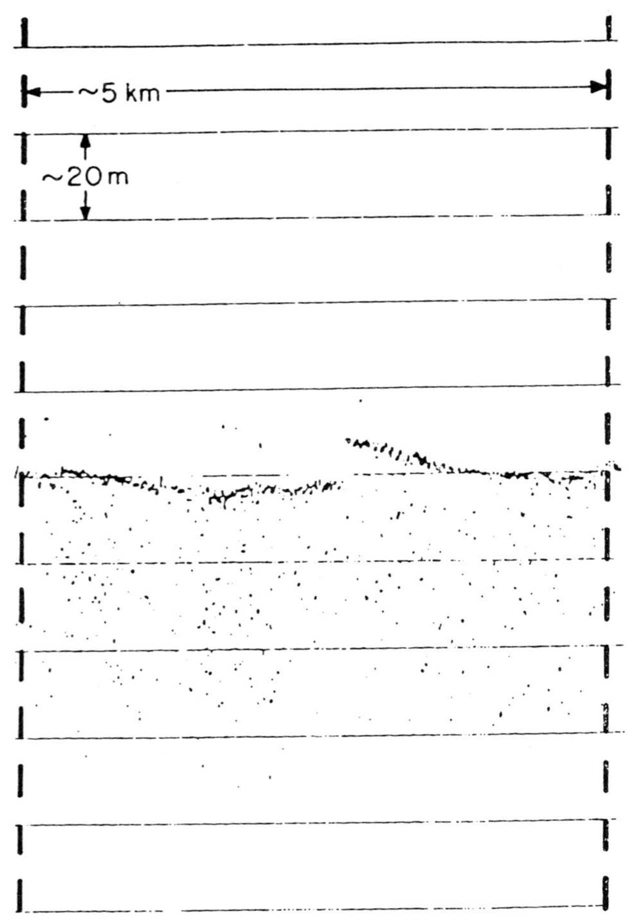

Fig. 11. Example of a 3.5-kHz bathymetric profile in the Victoria Land basin (see Figure 9), collected by R/V *S.P. Lee*, interpreted as a fault scarp cutting the sea bottom. Because this area was covered by grounded moving ice possibly as recently as about 7000 years B.P., the scarp must be Holocene in age.

between reflection soundings. In the McMurdo, northern Victoria Land, and Pensacola Mountains areas of the Transantarctic Mountains and in the Ross Sea and Ross Ice Shelf areas more accurate data have been gradually obtained but coverage is still very sparse. In this section we briefly discuss free-air anomalies over Antarctica (Figure 12) and the available Bouguer anomaly data for West Antarctica (Figure 13).

Free-Air Anomalies

Bentley [1968] compiled free-air gravity data (Figure 12) over 2° × 2° (222 km × 222 km) mean squares in Antarctica. He noted a mean free-air value near 0 for all of West Antarctica, implying isostatic balance on a regional basis for the entire area. Bentley discussed the –20 to –50 mGal "Transantarctic gravity anomaly" parallel to the Transantarctic

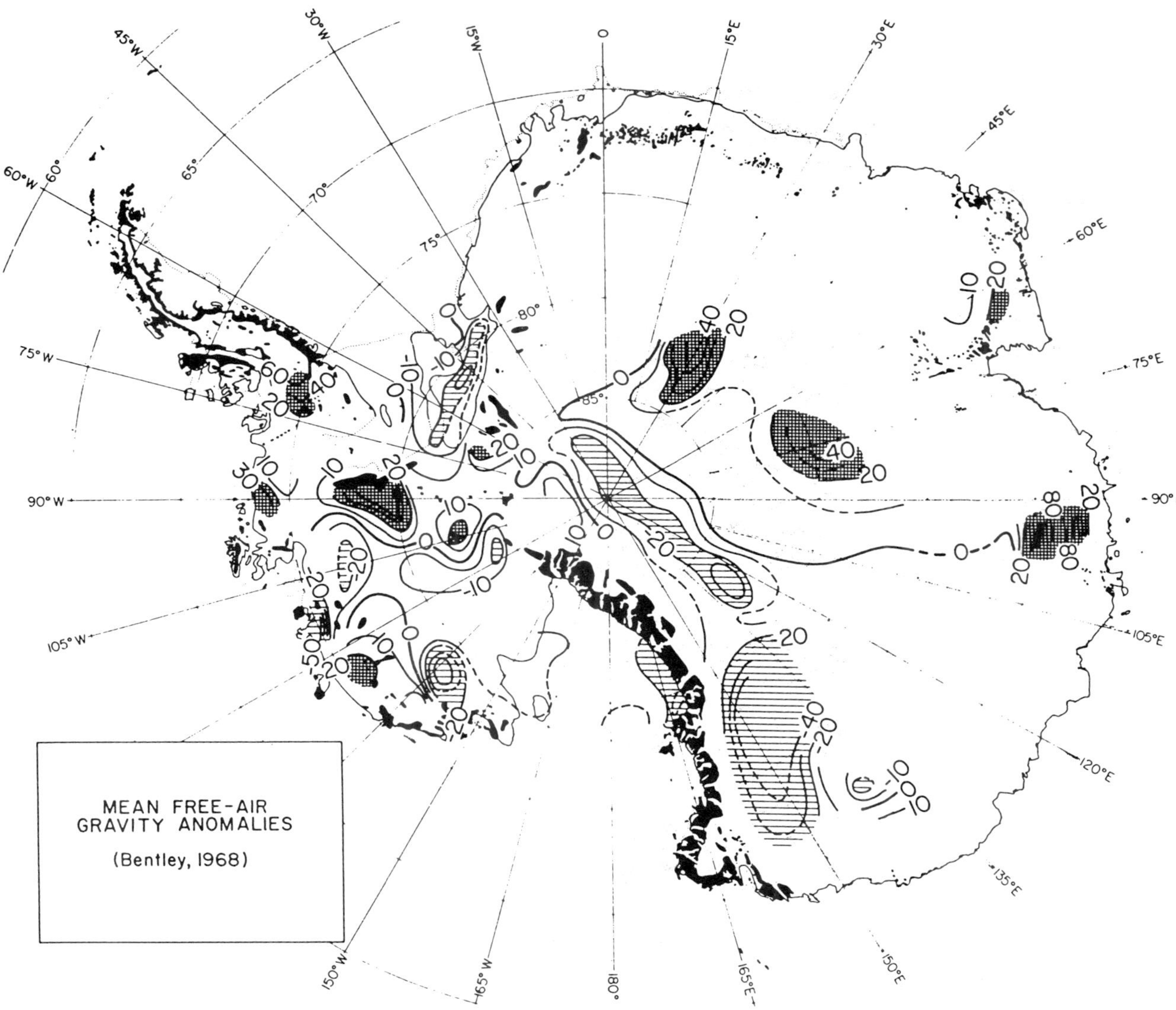

Fig. 12. Free-air gravity anomalies averaged over 2° × 2° squares from *Bentley* [1983], simplified from *Bentley* [1968]. Anomalies greater than 20 mGal in amplitude are indicated by horizontal hachures for negative anomalies and by cross-hatching for positive anomalies.

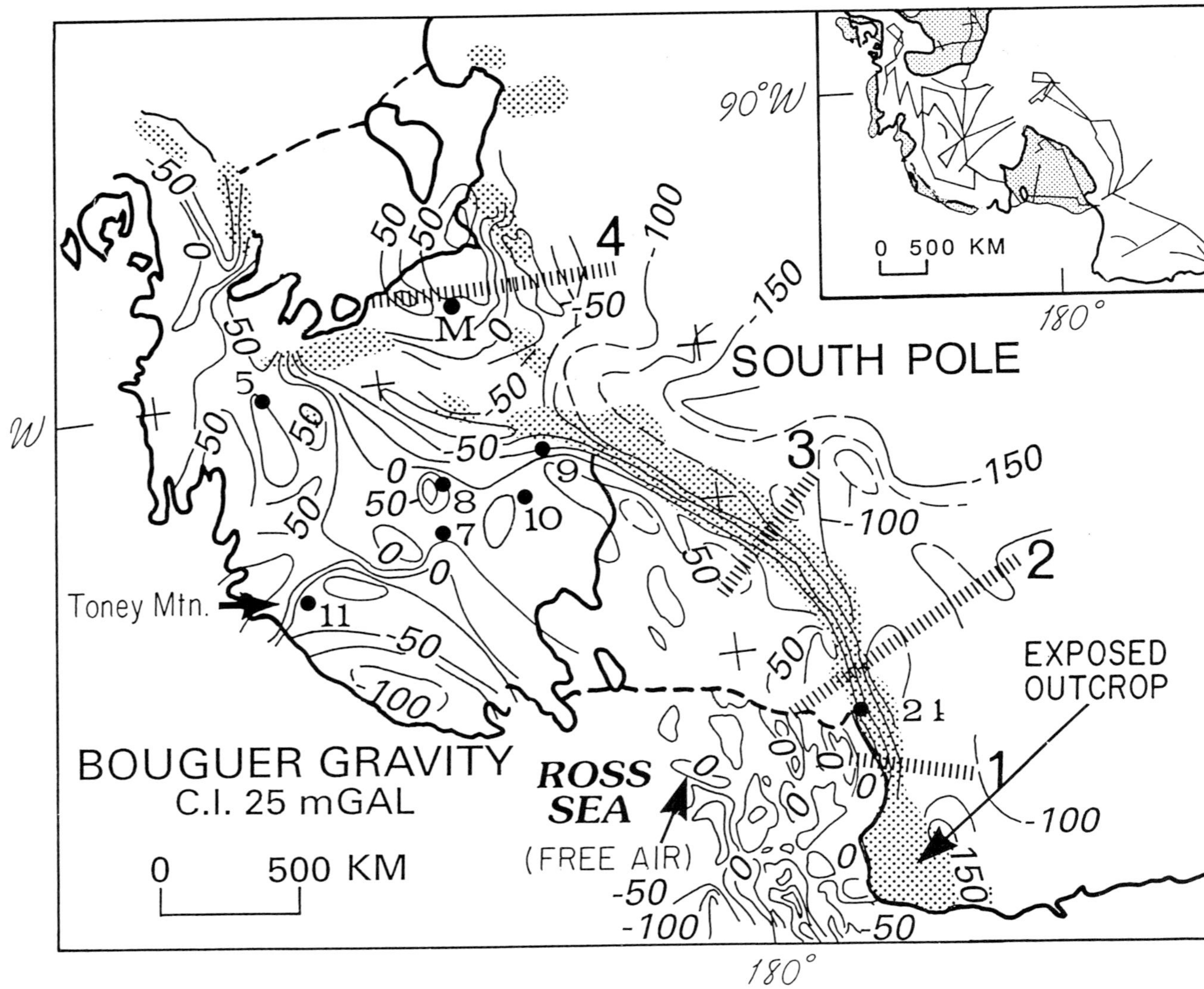

Fig. 13. Bouguer anomaly contour map for West Antarctica compiled from data collected at seismic reflection stations (about 30–40 km spaced) where ice thickness measurements were made by oversnow traverse parties led from 1956 to 1964 by J. C. Behrendt, C. R. Bentley, F. K. Chang, A. P. Crary, M. Hochstein, E. S. Robinson, E. Thiel, and F. Van der Hoeven. The inset map shows locations of these oversnow traverses. Additional data: Ross Ice Shelf from the map of *Robertson et al.* [1982]. The free-air anomaly map over the Ross Sea shelf is from *Davey and Cooper* [1987]. The heavy hachures indicate areas where data density allowed reasonably accurate calculation of gradients: 1 [*Duerbaum et al.,* 1989], 2 [*Robinson,* 1964*a*; *Smithson,* 1972; *Robinson and Splettstoesser,* 1984], *3* [*Robinson,* 1964*a*; *Robinson and Splettstoesser,* 1984], and 4 [*Behrendt et al.,* 1974]. Seismic velocity columns (Figure 16) from *Bentley and Clough* [1972] are indicated: 11 (at Toney Mountain), 7 (Byrd Station), 8 (Whitmore Mountains), 5 (Ellsworth Mountains), 24 (Ferrar dolerite near McMurdo), and 10 and 9 (Horlick Mountains). M is Moho reflection from M. Hochstein, 1963–1964 (cited by *Bentley* [1973]).

Mountains on the East Antarctic side which extends from northern Victoria Land nearly to the Pensacola Mountains. *Bentley* [1983] attributes the source of this anomaly to large glacio-isostatic imbalance, to abnormally thick or low-density crust, or to a deep-seated cause in the upper mantle. Although low-density sedimentary rock in the Wilkes basin (Figure 2) could partly explain the gravity low, the interpretation of a thick (>600 km) lithosphere in East Antarctica [*Stern and ten Brink,* 1989], resulting in a flexurally controlled basin which is regionally compensated, is largely the cause. Based on flexural rigidity and gravity models, the front of the Transantarctic Mountains in the McMurdo area is postulated to have a much higher rigidity (10^{25} N m) for the East Antarctic "cantilevered lithospheric plate" than for the less rigid (4×10^{22} N m), stretched, hot lithospheric plate underlying the Ross Embayment [*Stern and ten Brink,* 1989]. Thus, the Transantarctic (free-air) negative gravity anomaly (Figure 4) is evidence for regional compensation of the load of the Transantarctic Mountains as modeled by *Stern and ten Brink* [1989].

Free-air anomalies are always locally positive in mountains because the compensating mass, irrespective of thickness or rigidity of the lithosphere, is at a greater distance below the gravity station than the rugged mountain topography, so positive free-air anomalies over mountains in Figure 12 are expected. This is apparent in the data from the oversnow traverses near the Ellsworth Mountains and the southern Antarctic Peninsula [*Behrendt,* 1964*b*; *Behrendt et al.,* 1974]. A more recent paper illustrating this point [*Simpson et al.,* 1986] discusses isostatic anomalies in the United States and notes that "large amplitude anomalies can be produced by crustal bodies in complete local isostatic equilibrium." The map in Figure 12 is biased by the high topography, which therefore produces positive mean free-air anomalies over the 2° squares in the southern Antarctic Peninsula [*Behrendt,* 1964*b*], Pensacola and Ellsworth Mountains area [*Behrendt et al.,* 1974].

Bouguer Anomalies

The regional Bouguer anomaly map (Figure 13) of the West Antarctic rift system and Transantarctic Mountains area is more useful for our purposes than the free-air anomaly map. Figure 13 is a compilation of Bouguer anomalies from various sources [*Robinson,* 1964*a*; *Behrendt,* 1964*b*; *Behrendt et al.,* 1974; *Robertson et al.,* 1982; *Robinson and Splettstoesser,* 1984]. Data from 19 U.S. Geological Survey gravity stations acquired in the Beardmore Glacier area (collected by M. Hower and G. Perasso in 1985–1986) were used. We also included all the original 1957–1964 U.S. oversnow traverse data (used by *Bentley* [1968], who averaged the data over 2° × 2° squares and therefore smoothed the gradients over the rift flanks).

In using the Bouguer anomaly map, errors due to uncertainties in elevation (as great as ±50 m for the oversnow traverse data) and terrain effects must be kept in mind. The data in the mountains have the best elevation control but are subject to the greatest terrain effects. For example, the approximately 400 stations in the Pensacola Mountains area [*Behrendt et al.,* 1974] are estimated to have ±25 m absolute elevation error, corresponding to about ±5 mGal Bouguer anomaly error for stations on bedrock. Terrain corrections in the mountains are the most difficult to evaluate and could be as great as several tens of milligals. However, for most stations, terrain effects probably are 10 mGal or less [*Behrendt et al.,* 1974]. Terrain corrections for subglacial bedrock topography are not possible for most of the data used to compile the Bouguer anomaly map because of the scale and contour interval available for most maps (1/250,000 scale with 200-m contour intervals). Terrain corrections over the ice sheet are not possible with existing subglacial bedrock elevation data but could be as great as ±20 mGal over rugged subice topography buried by 1 km of ice [*Behrendt,* 1964*b*]. Most of the ice over the area depicted in Figure 13 is about 3 km thick, so probably the terrain errors are mostly less than 10 mGal here as well.

Considering all of the possible errors, we contoured the Bouguer anomaly data at a conservative 25-mGal interval and will only discuss the large regional anomalies ranging from 100 to 200 mGal in magnitude. Although the lack of data in the Marie Byrd Land–Ellsworth-Whitmore Mountains area is apparent, the map is nevertheless informative regarding the tectonics of the West Antarctic rift system.

Bouguer anomalies provide the primary evidence for the thinning of the crust throughout the ice-covered Byrd Subglacial Basin area of the West Antarctic rift system. Early crustal thickness maps based on Bouguer anomalies showed approximately 30-km-thick crust over the area of the West Antarctica rift in contrast with the approximately 40-km-thick crust in East Antarctica [*Bentley et al.,* 1960; *Woollard,* 1962; *Groushinsky and Sazhina,* 1982], calculated from the difference in Bouguer anomaly between these two areas. However, these regional studies were based on comparisons with other parts of the world and contributed information similar to that obtained by averaging regional elevations. There are no seismic refraction or reflection measurements of Moho depth anywhere in the ice-covered Byrd Subglacial Basin area. In this paper we interpret the data to indicate that the crust beneath the rift (Byrd Subglacial Basin–Ross Embayment) is probably closer to 20 km thick by adjusting the *Bentley et al.* [1960] and *Woollard* [1962] regional interpretation tied to the seismic determinations of 17–21 km depth to the Moho beneath the Ross Sea continental shelf. (See the seismic results below.) This value is typical for rift stage crust [e.g., *Klitgord et al.,* 1988].

Rift Shoulder Anomaly

One of the largest-magnitude Bouguer anomalies in the world was recognized as early as 1962 [*Woollard*, 1962] as marking the transition from West Antarctica to East Antarctica in the Ross Embayment (Figure 13). Bouguer anomalies increase by about 200 mGal from about −150 mGal over the Transantarctic Mountains to about +50 mGal over the inferred extended crust beneath the Ross Ice Shelf. We did not compute Bouguer anomalies over the Ross Sea shelf but include the free-air anomaly map of *Davey and Cooper* [1987] for this area. For a mean water depth of 500 m, a Bouguer correction of +34 mGal would need to be added to the regional free-air anomaly field over the Ross Sea shelf. Because water depths vary from 400 to 1400 m (e.g., Bouguer corrections of 28 to 96 mGal) over the shelf, caution should be used in using the free-air map. *Davey and Cooper* [1987] used bathymetry and seismic reflection data along specific free-air anomaly profiles in their interpretations.

Many authors [*Robinson*, 1964*a*; *Smithson*, 1972; *Behrendt et al.*, 1974; *Robinson and Splettstoesser*, 1984; *Davey and Cooper*, 1987; *Stern and ten Brink*, 1989] have computed theoretical models and explained the great change in gravity between West Antarctica and East Antarctica as a result primarily of an abrupt change in crustal thickness across the front of the Transantarctic Mountains. There are significant regional differences, however, as revealed by these models along the Transantarctic Mountains.

If we examine the 200-mGal change at the front of the Transantarctic Mountains near McMurdo and assume, as a first approximation, an infinite slab having a density contrast of 0.4 g/cm^3 at the crust-mantle boundary to account for the entire change, then the change in crustal thickness, using 41.85 mGal/km for a density contrast of 1.0 g/cm^3, $\Delta h = 200/(0.4 \times 41.85) = 12$ km. Were the 200-mGal difference to be caused solely by a difference of crustal thickness as great as 20 km, a density contrast of 0.24 g/cm^3 would be implied.

This value is consistent with those calculated by *Robinson* [1964*a*] and *Robinson and Splettstoesser* [1984]. There is a significant difference between the maximum 200-mGal change across the front of the Transantarctic Mountains near McMurdo (profile 2 in Figure 13) and the maximum 130-mGal anomaly range at the front of the Transantarctic Mountains in the Pensacola Mountains area.

Bouguer Anomaly Gradient Across Transantarctic Mountain Front

We examined the steepest gravity gradients measured across outcrops of Paleozoic or older sedimentary or crystalline rock at the front of the Transantarctic Mountains. In the McMurdo area, the Bouguer anomaly gradient ranges from 4 to 7 mGal/km and, based on fewer data, appears to be as steep as 4 mGal/km in the Beardmore Glacier area [*Robinson*, 1964*a*; *Robinson and Splettstoesser*, 1984]. Two-dimensional gravity models have been computed across the front of the Transantarctic Mountains in the Beardmore Glacier and McMurdo areas of the Ross Embayment; two in the McMurdo area are shown in Figures 14*b* and 14*c* along profile 2 in Figure 13. A common feature of these models (e.g., Figure 14*c*) is a steep or near-vertical 10- to 20-km step in the Moho, interpreted generally as a fault extending to the surface at the front of the Transantarctic Mountains.

In contrast, the steepest gradient in the Pensacola Mountains area at the Filchner Ice Shelf is only about 2 mGal/km (Figure 14*d*) [*Behrendt et al.*, 1974]. The lower 2-mGal/km gradient observed across the 130-mGal gravity change in the Pensacola Mountains section of the Transantarctic Mountains was fit with an approximate 15°–20° dip on the assumed Moho [*Behrendt et al.*, 1974]. The model is tied to a 24-km depth to the Moho on the West Antarctica side of the Pensacola Mountains determined by wide-angle seismic reflection (Figure 14). In this Pensacola Mountains profile, a crustal thickness increase of only about 8 km was required across the front of the Transantarctic Mountains to account for the observed gravity difference. One explanation for this difference is that the West Antarctic rift system trend, which is superimposed in the Ross Embayment area on the Jurassic Transantarctic rift, diverges at the inland end of the Ross Ice Shelf and continues along the Ellsworth-Whitmore Mountains trend.

Another possibly more likely explanation for the lower 2-mGal/km Bouguer anomaly gradient in the Pensacola Mountains is the deliberate omission of the Dufek intrusion in the location of the profile selected for modeling [*Behrendt et al.*, 1974]. Although it was not possible to measure a comparable gravity gradient for stations on the exposures of the Dufek intrusion because of variable erosion and terrain effects, *Behrendt et al.* [1974] calculated about 85-mGal maximum gravity effect of the thickest section of gabbro. For a density contrast of 0.27–0.33 g/cm^3, 85 mGal corresponds to 8.8 to 6.2 km total thickness of the Jurassic tholeiitic intrusion. We suggest that possibly the exposed Dufek intrusion is an example of underplated and shallow- to lower-crustal intrusions which we infer to exist elsewhere along the Transantarctic Mountain front, where gradients of 4–7 mGal are observed. If the Dufek intrusion was buried by several kilometers of sedimentary rock, the high gradient would result.

We examined maximum possible gravity gradients in the Ross Embayment area with a simple extreme model (Figure 14*a*), again assuming a density differential of 0.4 g/cm^3 across the Moho to explain the gravity difference. Various depths to the Moho determined by seismic investigations are reported beneath the Ross Sea shelf: at McMurdo Sound (21 km) [*McGinnis et al.*, 1985], at Central basin (17–21 km) [*Trehu et al.*, 1989], and near the coast beneath the Transantarctic Mountains at about 74°34′S (22 km) [*O'Connell et al.*, 1989] (Figure 15). If we assume a 20-km Moho depth on the Ross Sea side of the Transantarctic Mountains with a vertical step to 40 km at the front of the mountains, we obtain a gradient of 4 mGal/km significantly lower than the maximum of 7 mGal/km in the McMurdo area reported by *Smithson* [1972] (Figure 14*c*). This model also produces an unreasonable maximum difference of 334 mGal for an infinite slab compared with about 200 mGal observed. Therefore we conclude (as did *Smithson* [1972]) that a significant excess mass, such as a high-density intrusion like the Dufek intrusion, exists at a shallow depth within the crust on the West Antarctic side of the Transantarctic Mountains. Because of the steep gradient observed, contrast in mantle density as great as 0.4 g/km^3 could not account for the anomaly. *Robinson and Splettstoesser* [1984] (Figure 14*b*) also required a high-density (3.0 g/cm^3) layer within the crust to fit the observed gradient in the McMurdo area. These results are consistent with gravity models for marine data in the Ross Sea [*Davey and Cooper*, 1987], although such models are

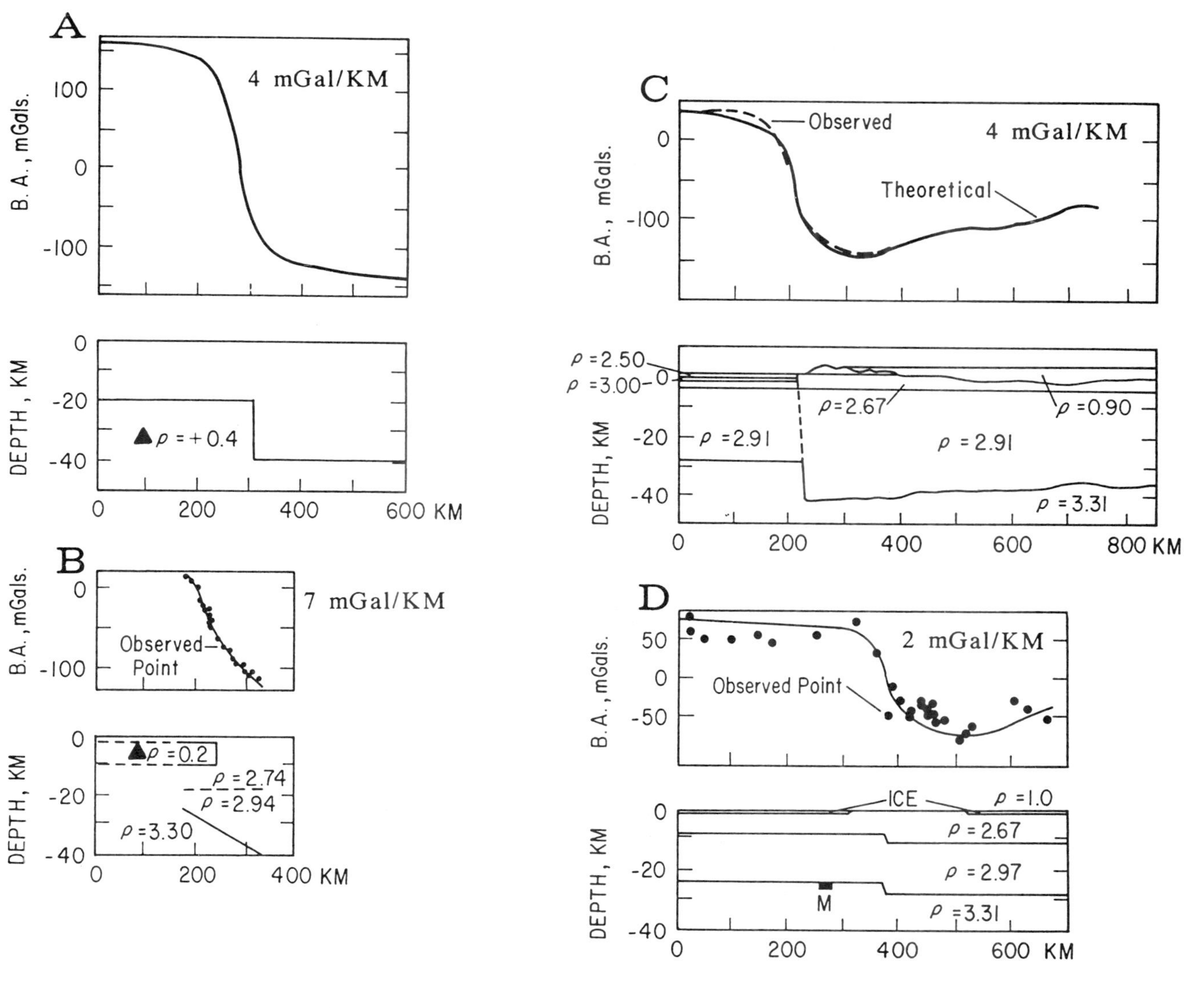

Fig. 14. Theoretical and observed Bouguer anomaly profiles for several models with maximum gradients indicated. (*a*) Model for an assumed crustal thickness change of 20 km from 20 to 40 km with a simple density contrast of 0.4 g/cm^3 across an assumed Moho. The maximum gradient is 4 mGal/km and the total anomaly range is about 330 mGal. This type of model cannot explain the observed data in the profiles shown in Figures 14*b*–14*d* and Figure 13 but is required to produce the steep gravity gradient solely by a change in crustal thickness. (*b*) Model approximately along 2 in Figure 13 [*Smithson*, 1972]. (*c*) Also approximately along 2 in Figure 13 [*Robinson*, 1964*a*, *b*; *Robinson and Splettstoesser*, 1984]. (*d*) Approximately along 4 in Figure 13 across the Pensacola Mountains [*Behrendt et al.*, 1974]. M is Moho reflector from M. Hochstein, 1964 (cited by *Bentley* [1973]).

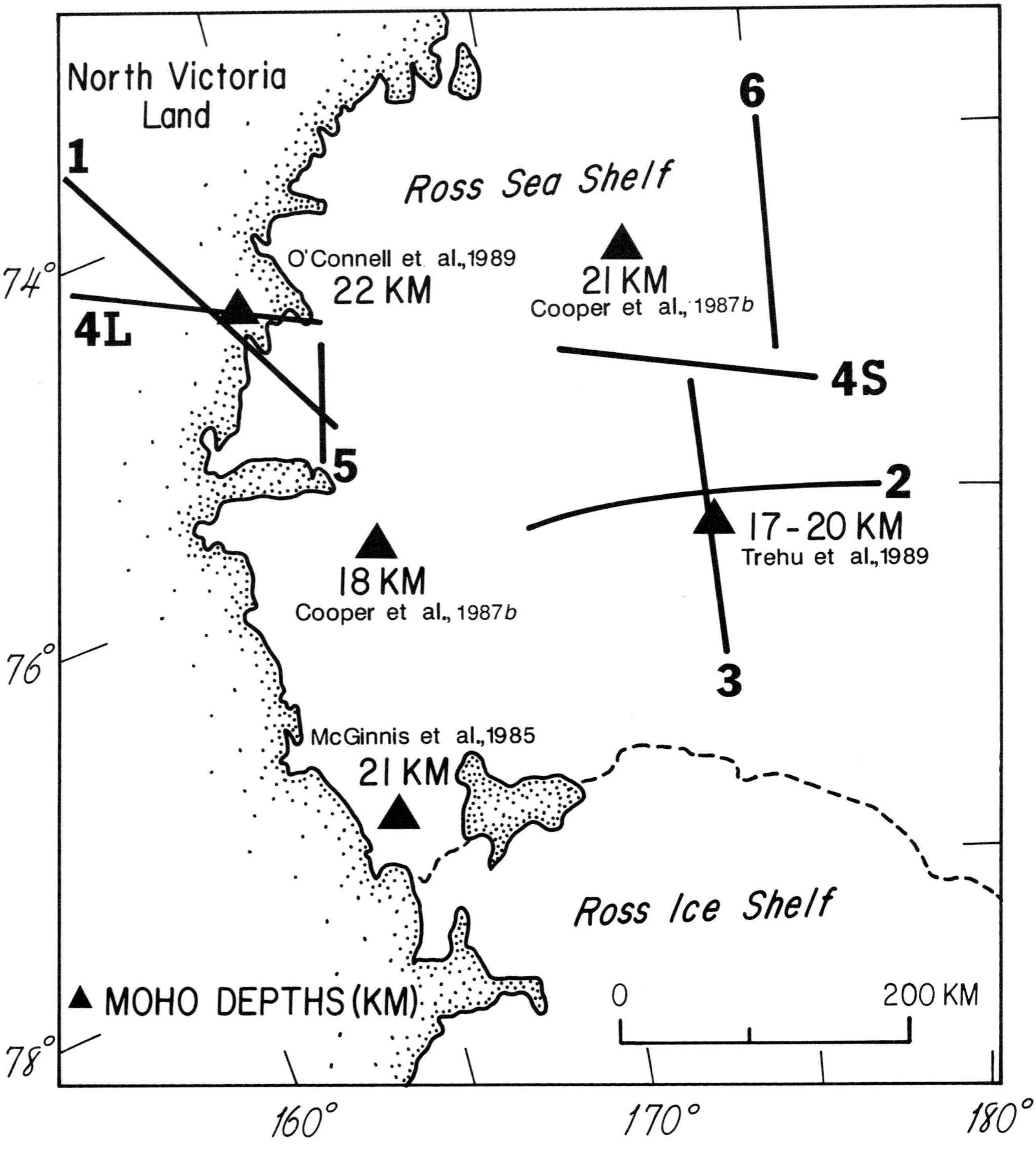

Fig. 15. Locations of large offset seismic experiments along profiles in the Ross Sea shelf–northern Victoria Land area. Profiles 1–6 are from the 1988–1989 GANOVEX V expedition. Profile 3 is along the strike of the Central basin (see Figure 7).

complicated by the low-density Cenozoic sedimentary rocks filling the rift basins beneath the Ross Sea shelf. The marine profiles, of course, did not cross the Transantarctic Mountains rift shoulder but did match part of the steep gradient. We discuss seismic evidence for high-density rock at the East Antarctica–West Antarctica transition in the next section.

We can conclude from the sparse Bouguer anomaly data (Figure 13) across the Transantarctic Mountains that the significant difference in the observed Bouguer anomaly range (about 200 mGal in the Ross Embayment and 130 mGal in the Pensacola Mountains–Filchner Ice Shelf area) and the difference in gradient (about 4–7 mGal/km and 2 mGal/km, respectively) indicate that greater crustal thinning and a notable excess mass are present at this boundary beneath the Ross Embayment. We interpret the excess mass to be associated with shallow mafic intrusions within the attenuated crust in the West Antarctic rift system.

Victoria Land Basin

Important questions remain about the possible presence of underplated or shallow-crustal high-density intrusions suggested by the 4- to 7-mGal/km gradients or thick sections of basalt within the rift basins. The high-density basin fill (2.7 g/cm^3) used below 5–7 km to fit the gravity anomaly over the 14-km-thick section in the Victoria Land basin [*Davey and Cooper*, 1987; *Cooper et al.*, 1987*a*] suggests that a substantial part of the section could be high-density basalt flows contrasted to lower-density sedimentary rock, possibly similar to those observed in the Midcontinent rift system of North America [*Behrendt et al.*, 1990]. The high-density mass beneath the Ross Sea basins was also shown by *Hayes and Davey* [1975] and inferred in the Ross Embayment [*Cooper et al.*, 1991]. We consider this supportive of the unwarped Curie isotherm hypothesis for the negative magnetic anomaly discussed previously; i.e., high-density basalt could have a low magnetization because of shallow high temperatures. The high seismic velocities, 5.6 km/s at 6.8 km depth within parts of the basin [*Cooper et al.*, 1987*b*], also support the basalt interpretation as discussed below.

Marie Byrd Land Area

The regional Bouguer anomaly gradient associated with the front of the Transantarctic Mountains in the Ross Embayment continues along the West Antarctic rift system shoulder to the Ellsworth Mountains (Figure 13). The Bouguer anomalies range from a minimum observed value of about −75 mGal in the Whitmore Mountains (not contoured because the data are too sparse) to about +50 mGal over the Byrd Subglacial Basin. We know from the combination of magnetic depth estimates discussed above and the limited seismic refraction data discussed in the next section that normal basement density rocks are probably quite shallow (<1 km) beneath the ice in the Byrd Subglacial Basin. Because there are no gravity data on the outcrops of the Ellsworth Mountains, we cannot determine maximum range and gradient there.

The only places along the West Antarctic rift system rift shoulder where gradients equal to or steeper than 4 mGal/km have been measured are where the Cenozoic and Jurassic magmatic rock exposures are approximately coincident. Along the Ellsworth-Whitmore-Horlick Mountains section of the Cenozoic rift shoulder no exposures of either Jurassic tholeiites or late Cenozoic volcanics are known.

The Bouguer values across the Amundsen-Bellingshausen flank of the West Antarctic rift system decrease to about −100 mGal from the +50 mGal Bouguer anomaly over the Byrd Subglacial Basin. The gradient does not appear as steep over this flank of the rift as across the Ellsworth-Whitmore side of the rift, which again may demonstrate the asymmetry of the West Antarctic rift system, although caution should be taken when inferring gradients in areas of widely spaced data. The 150-mGal difference corresponds to 9-km difference in depth to the Moho rather than the 12 km calculated above for the 200-mGal difference across the rift shoulder, assuming the same 0.4 mGal/km across the crust-mantle boundary only.

Southern Antarctic Peninsula–Ellsworth Mountains Area

Steep gradients of 100 mGal or greater exist in the southern Antarctic Peninsula–Ronne Ice Shelf area (Figure 13). The maximum range is from about −70 to +60 mGal [*Behrendt*, 1964*b*]. Minimum values are probably not as negative over the narrow southern Antarctic Peninsula because of the geometric effect of the high topography on the gravity field as pointed out for the two isostatic models calculated by *Behrendt* [1964*b*]. As was discussed in the magnetics section, we are uncertain where the West Antarctic rift system terminates toward the Ronne Ice Shelf. *Garrett et al.* [1987] and *Kellogg and Rowley* [1989] discussed evidence for Cenozoic extension between the Antarctic Peninsula and the Ellsworth Mountains and right-lateral strike-slip faulting (the Weddell fault system) along the east side of the Antarctic Peninsula. The Ellsworth fault system [*Kellogg and Rowley*, 1989] lies within the complex topography separating the Antarctic Peninsula and the Ellsworth Mountains and forms a boundary with the area of extended crust suggested by gravity data (Figure 13) and the linear prominent (>1000 nT) magnetic anomalies (Figure 2).

SEISMIC INVESTIGATIONS

Large Offset Surveys—Ice Sheet

Multichannel seismic reflection data (analog) collected on the oversnow traverses in the area shown by the Bouguer anomaly data in Figure 13 were primarily made to determine ice thickness; few reflections from deeper than the ice bed were reported. The only velocity control, used to interpret vertical incidence reflection profiles, is from large offset (reflection and refraction) surveys. Refraction results obtained for the upper crust at about 33 sites in the area of Figure 2 have been summarized [*Bentley and Clough,* 1972; *Rooney et al.,* 1987]. All data were collected on the ice sheet, which has a seismic velocity of 3.9 km/s. Therefore evidence of lower-velocity sedimentary rock would not have been observed as a first arrival, although it was inferred indirectly in several cases [*Bentley and Clough,* 1972; *Behrendt et al.,* 1974; *Rooney et al.,* 1987].

Marie Byrd Land–Horlick Mountains Area

Figure 16 shows a seismic velocity column across the West Antarctic rift system; the locations of stations (11, 7, 10, and 9) along the profile are indicated in Figure 13. The 4.5-km/s velocity measured at Toney Mountain can unambiguously be associated with andesitic-basaltic flows exposed nearby (Figure 2). In the same profile the 6.1-km/s layer is interpreted as high-grade metamorphic(?) basement. In the Byrd Station profile the 4.3-km/s velocity is also correlated with magnetic volcanic rock [*Behrendt and Wold,* 1963], on the basis of the magnetic interpretations discussed previously. The distinct velocity change [*Bentley and Clough,* 1972] on either side of the axis of the Byrd Subglacial Basin (compare columns 7 and 10 in Figure 16) can be seen. The break in the magnetic anomaly pattern in this area (Figures 5 and 6) discussed in the magnetics section also occurs between these two profiles. Seismic refraction data [*Bentley and Ostenso,* 1961; *Bentley and Clough,* 1972] show about 1 km of 5.2–5.3 km/s velocity overlying 6.3-km/s and 6.1-km/s rock, respectively, in the area interpreted as thick nonmagnetic rock (east of break in Figures 5 and 6). These velocities could indicate granitic terrane as originally suggested [*Bentley and Ostenso,* 1961], on the basis of the nonmagnetic nature of the subice rock; however, as one of these refraction determinations (profile 5 in Figure 13) is close to the Ellsworth Mountains, we still subscribe to the interpretation of *Behrendt and Wold* [1963] that the velocities correspond to the thick metasedimentary rock section (lower Paleozoic) exposed in the Ellsworth Mountains.

Profile 10 in Figure 16 near the Horlick Mountains is on the Cenozoic West Antarctic rift shoulder, as defined in this paper. The 6.7-km/s and 7.0-km/s velocities are more typical of lower crust [*Bentley and Clough,* 1972] and are similar to others reported in the area away from profile C—C′ in Figure 16. These high velocities are also similar to those measured in McMurdo Sound [*McGinnis et al.,* 1985; *Kim et al.,* 1986] and beneath the Ross Sea shelf [*Cooper et al.,* 1987*b*; *Trehu et al.,* 1989]. We interpret these high velocities, typical of lower crustal rocks, to be associated with mafic magma intruded during rifting. Because the Horlick Mountains are located where the Jurassic Transantarctic rift and Cenozoic West Antarctic rift shoulder are coincident, however, the high seismic velocities could be the result of either Early Jurassic or Cenozoic intrusions or both. A velocity of 6.8 km/s (profile 24 in Figure 13) [*Bentley and Clough,* 1972; *Robinson and Splettstoesser,* 1984] was reported for the Jurassic Ferrar dolerite near McMurdo, whereas a velocity of 4–5 km/s is typical of a Cenozoic basalt (as measured, for example, at Toney Mountain [see *Bentley and Clough,* 1972]). However, profile 8 in Figure 13 near the Whitmore

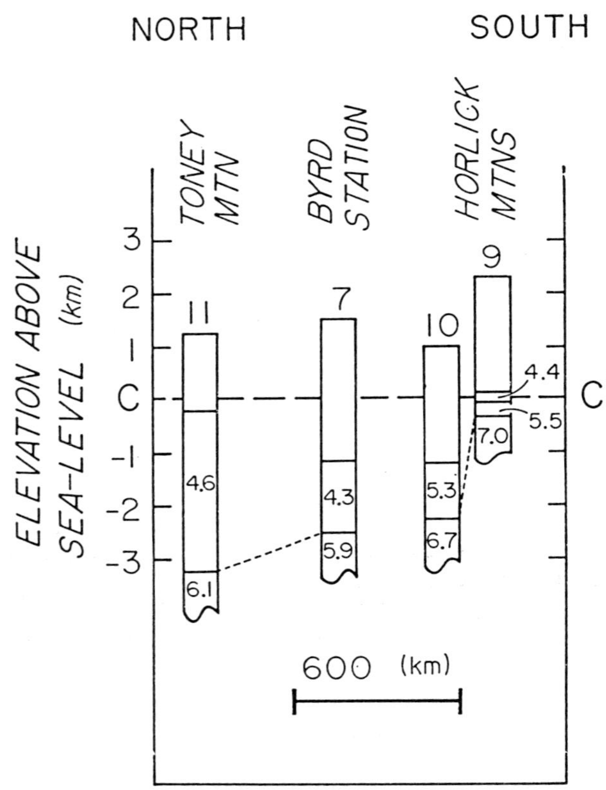

Fig. 16. Seismic velocity columns 11 to 9 along the section (see Figure 13), measured from the 1956–1962 oversnow traverses (section C—C′), from *Bentley and Clough* [1972]. Velocities are shown in kilometers per second. Correlations between columns are shown by dashed lines. The blank upper part of each column corresponds to ice V_p = 3.9 km/s. We interpret the velocities as follows. The 4.6-km/s velocity at profile 11 (at Toney Mountain) is correlated with exposed volcanic rock. The 4.3-km/s velocity at Bryd Station (profile 7) is also correlated with volcanic rock because of shallow sources of magnetic anomalies in the area. At profile 9 (near the Horlick Mountains) (see Figure 2), the 4.4-km/s velocity layer is probably Beacon Supergroup Paleozoic-Mesozoic sedimentary rock; the 6.7-km/s velocity at profile 10 and 7.0-km/s velocity at profile 9 are interpreted as mafic or ultramafic intrusive rock. The 5.9–6.1-km/s velocities at profiles 7 and 11 are probably crystalline granitic basement, as are possibly the lower 5.3-km/s and 5.5-km/s velocities at profiles 10 and 9.

Mountains (about 100 km on the rift side) exhibits an approximate 7 km/s velocity about 4 km below sea level underlying 5.8-km/s basement [*Bentley and Clough*, 1972]. This area overlies the interpreted rift well away from the magmatic rocks of the Jurassic Transantarctic rift system, suggesting that Cenozoic mid- and lower-crustal mafic intrusions may be the source of the 7 km/s velocity there. In contrast to these high shallow or midcrustal velocities is the comparatively low 6 km/s mean (subice) crustal velocity measured by M. Hochstein [*Behrendt et al.*, 1974], interpreted as evidence of a thick section of low-density nonmagnetic sedimentary rock at the Moho reflection location (Figure 13) northwest of the Pensacola Mountains. No high-density mass was required within the crust there to fit a model to the observed Bouguer anomaly profile (Figure 14*d*).

McMurdo Sound

McGinnis et al. [1985] and *Kim et al.* [1986] discussed deep reflection and refraction profiles in McMurdo Sound that involved using land techniques on sea ice. *McGinnis et al.* [1985] observed a 6.5-km/s refractor as shallow as 5.4 km beneath a 5.0-km/s refractor on a north-south profile. On another east-west profile they report a 7.2-km/s refractor in the lower crust. The velocity below the Mohorovic discontinuity, which occurs at 21 km depth, is 8.2 km/s (Figure 15). Their 12-fold reflection profile shows bands of reflections in the lower crust similar to those interpreted elsewhere as caused by underplated basalt associated with rifting [e.g., *Behrendt et al.*, 1988, 1989]. A lower crustal velocity of 7.2 km/s is reasonable for stretched underplated and intruded lower crust; *Furlong and Fountain* [1986] discuss examples of such crust having seismic velocities of 7.0–7.8 km/s. *McGinnis et al.* [1985] assume a 2.94-g/cm^3 density for the 6.5-km/s layer in a gravity model that is seaward of the steepest part of the gradient measured by *Smithson* [1972]. Therefore we consider this density and velocity consistent with our previous discussion of gravity data. *Kim et al.* [1986] show bands of reflections and diffractions as late as about 7.8 s beneath the Victoria Land basin and suggest these are sideswipe, but in appearance they are quite similar to reflections from probable underplated material reported for other rifts [e.g., *Behrendt et al.*, 1988, 1990].

Marine Large Offset Surveys—Ross Sea

In striking contrast in number to the sparse large offset seismic surveys done over ice sheets (only about 30–40 profiles on "land" in 33 years), marine data are several orders of magnitude more efficient to collect than land data. For example, the R/V *S.P. Lee* obtained 39 sonobuoy profiles over the Ross Sea shelf in a few weeks in 1984. Interpretations generally were made only to midcrustal depth, but some resulted in probable Moho reflections. *Cooper et al.* [1987*b*] discussed these results, and *Davey et al.* [1983] discussed 98 older, shallow-penetration sonobuoy profiles.

Velocities reported for the sedimentary rocks in the three basins beneath the Ross Sea shelf (Figures 7 and 17) ranged from 1.7 to 4 km/s in the upper few kilometers of the Victoria Land basin but increased to as great as 5.6 km/s in the lower 6–8 km of this 14-km-thick basin [*Cooper et al.*, 1987*a*, *b*]. Because these velocities are significantly higher than the 4.4 km/s measured for the Devonian to Jurassic Beacon Supergroup rocks exposed nearby in the Transantarctic Mountains and buried beneath 1–2 km of ice elsewhere in the Transantarctic Mountains (e.g., profile 10 in Figure 16), and considering the 4.6 km/s measured for Cenozoic volcanic rocks at Toney Mountain (Figure 16), the rocks are probably in fact rift-related volcanic rocks of uncertain age. Whether or not these layered [*Cooper et al.*, 1987*a*] rocks (in the deepest Victoria Land basin) are Cenozoic, they must predate the overlying sedimentary rocks which *Cooper et al.* [1987*a*] interpret to be as old as early Oligocene, from correlations with the CIROS-1 drill site (Figure 7) [*Barrett*, 1989] on the flank of the Victoria Land basin. Ages of the deep, high-velocity layered reflectors probably are from Late Mesozoic to Paleogene [*Cooper et al.*, 1987*a*]. Another hypothesis to explain the high velocities would be diagenetically altered or metamorphosed sedimentary rock. We think this is unlikely considering they are probably not older than the Devonian-Jurassic Beacon Supergroup rocks which have lower velocity (4.4 km/s) where measured [*Bentley and Clough*, 1972].

The highest crustal velocities measured in the Ross Sea [*Cooper et al.*, 1987*b*] are 7.3 km/s at 8.3 km depth beneath the east flank of the Victoria Land basin and 7.4 km/s at 3.0 km depth, 7.4 km/s at 3.5 km depth, and 6.9 km/s at 2.8 km beneath the Coulman high (Figure 7). *Cooper et al.* [1987*b*] reported a velocity of 6.5 km/s from within the central region of the Victoria Land basin at a depth of 9.5 km as acoustic basement on the reflection profiles. Although *Cooper et al.* [1987*b*] interpreted rocks with velocities greater than 6.3 km/s as igneous-metamorphic basement, here we refine this to infer that velocities of 6.9–7.4 km/s probably indicate mafic or ultramafic intrusions typical of underplated crust [e.g., *Furlong and Fountain*, 1986] at quite shallow depths beneath the Ross Sea. As mentioned in the gravity section, these rocks may be the source of the high densities required within the crust to fit the observed steep gravity gradients and a high range of total Bouguer gravity values and are additional evidence of rifted extended crust.

Moho Determinations—Ross Sea Shelf

Cooper et al. [1987*b*] report deep reflections at two sites which we interpret as the Moho because they have depths similar to other Moho determinations discussed below. The corresponding depths to Moho

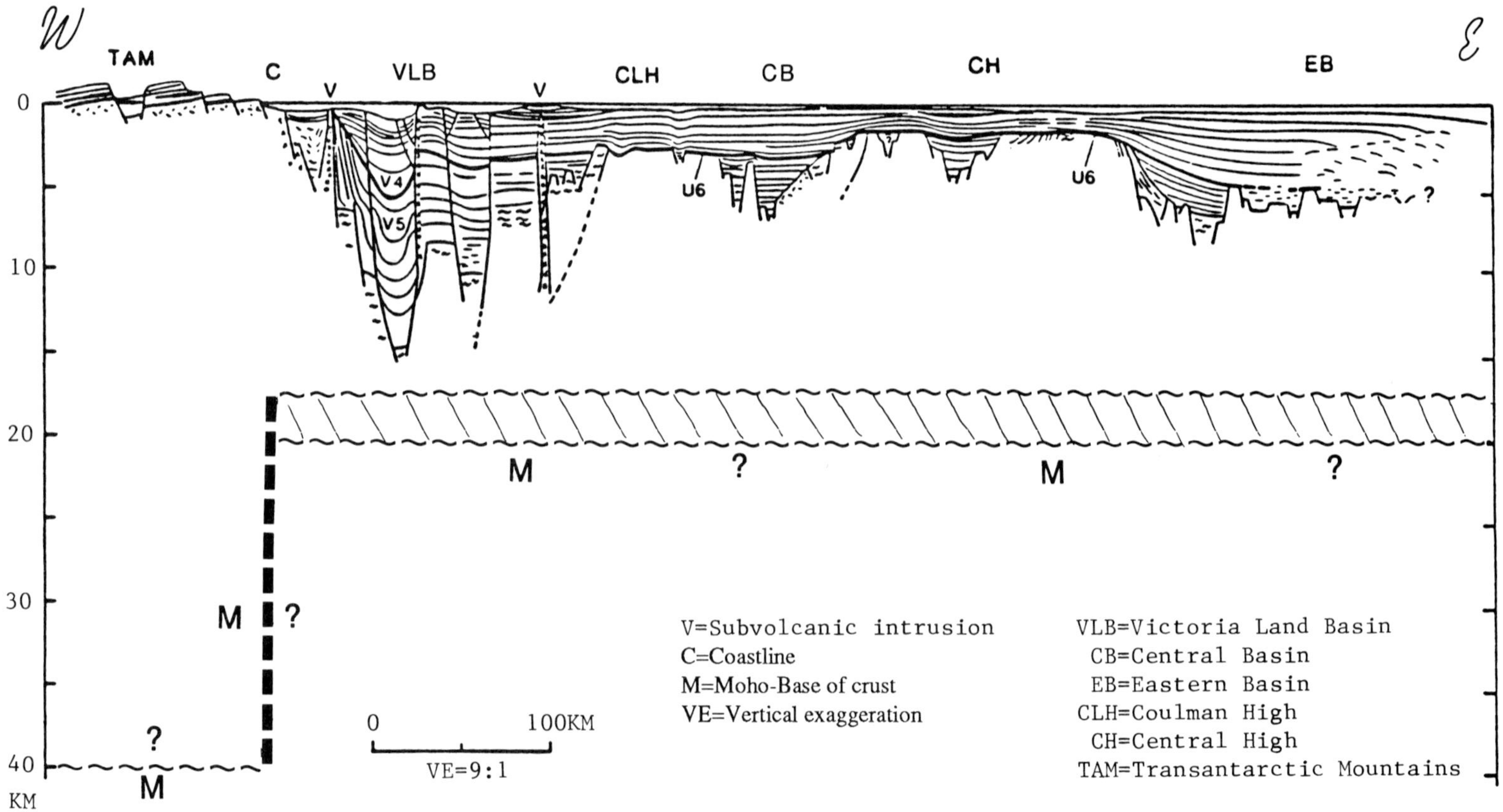

Fig. 17. Cartoon illustration of the geologic structure across the Ross Sea shelf, modified from the seismic reflection interpretation of *Cooper et al.* [1990] for the upper crust and based on preliminary seismic crustal thickness determinations from GANOVEX V (see Figure 15). Unconformity U6 [*Hinz and Block*, 1983] is correlated with that between acoustic units v4 and v5 [*Cooper et al.*, 1987*a*]. The Moho is shown as a horizontal band from 18 to 20 km to indicate uncertainty in its depth. The approximate location of this profile is indicated by W—E in Figure 7. The vertical dashed line indicates that the Moho probably changes thickness by up to 20 km at the rift shoulder (Transantarctic Mountains front); however, the dip of the Moho in the transition zone is unknown but is probably not vertical or even necessarily steep. Note that the old crust has been nearly rifted through beneath the Victoria Land basin.

are 18 km beneath the central Victoria Land basin and 21 km beneath the Coulman high (Figure 15).

BGR and USGS (GANOVEX V, 1984–1985) conducted a large offset seismic experiment on land and at sea using a 43.5-L air gun source (Figure 15). From these results *Trehu et al.* [1989] reported Moho depths of 17–21 km beneath the Central basin with a high-velocity lower crustal transition layer, and *O'Connell et al.* [1989] reported a Moho depth of 22 km on land near the coast. A similar depth to the Moho of 21 km beneath McMurdo Sound was reported by *McGinnis et al.* [1985]. All of these Moho depths are consistent with each other and approximately consistent with the earlier gravity models of *Davey and Cooper* [1987], which suggested attenuated (rifted) crust beneath the basins.

The 17–21 km depths to Moho beneath the Ross Sea shelf are typical of extended and rifted continental crust, as described and defined by *Klitgord et al.* [1988] beneath the east coast area of North America, and the greater than 7 km/s velocities reported in the lower crust in the Ross Embayment area are typical of transitional crust. The very shallow high velocities (6.9–7.4 km/s) beneath the continental shelf are unusual.

Crustal Thickness Beneath the Ross Embayment–Byrd Subglacial Basin

As discussed previously, the Bouguer anomaly map (Figure 13) indicates approximately the same range (0 to +50 mGal) throughout the interpreted extended crust beneath the West Antarctic rift system from the Ross Embayment to the Byrd Subglacial Basin. Therefore, by comparison with measured depths to the Moho beneath the Ross Sea shelf, we conclude that the Moho depths throughout this topographically low area of the rift are more likely closer to 20 km than to 30 km, as previously reported [e.g., *Bentley et al.*, 1960; *Woollard,* 1962]. The earlier interpretations did recognize the continuity of what we now interpret as extended rifted crust, from the Ross Ice Shelf through the Byrd Subglacial Basin to the Bellingshausen Sea. There is no evidence for oceanic crust beneath the Byrd Subglacial Basin.

The 24-km depth to Moho obtained in 1963 from large offset seismic results by M. Hochstein (Figures 13 and 14*d*) [*Behrendt et al.*, 1974] northwest of the Pensacola Mountains appears more reasonable for the low topography in that area of West Antarctica than was realized at the time. The question remains whether the possible extended crust (Figure 14*d*) bordering the Pensacola Mountains in the Weddell Sea–Filchner Ice Shelf is the result of Jurassic Transantarctic rifting or Cenozoic rifting. The absence of any magmatic rocks younger than Jurassic suggests the former, but the faulted uplift of the Transantarctic Mountains in this area requires tectonic activity postdating the Jurassic Dufek intrusion (but possibly as old as Jurassic).

Marine Multichannel Seismic Reflection Surveys

Since 1980, approximately 35,000 km of marine common-depth-point reflection profiles have been acquired in the Ross Sea (Figure 7); data from U.S. and German research have been described by *Hinz and Block* [1983], *Hinz and Kristofferson* [1987], *Cooper et al.* [1987*a*], and *Cooper et al.* [1991]. These authors identify up to seven acoustic units and unconformities within the reflection section, which they interpret to be of late Mesozoic and younger ages. Acoustic basement may be composed of metasedimentary and/or igneous units of Precambrian to late Mesozoic age. A schematic interpretation of these data along one profile across the three basins in the Ross Sea is shown in Figure 17.

The only control for age and lithology of interpreted reflectors offshore is provided from sparse drilling in the central (Deep Sea Drilling Project (DSDP)) and western (CIROS and McMurdo Sound Sedimentary and Tectonic Study (MSSTS)) Ross Sea (Figure 7) and is inferred from recycled and glacial-erratic materials found within the Ross Sea region (see summaries by *Hayes and Frakes* [1975], *Webb* [1981], *Truswell* [1983], *Barrett* [1986], *Davey* [1987], and *Barrett et al.* [1987, 1991]). Glacial-marine sedimentary rocks of early Oligocene and younger age comprise the upper part of the thick offshore sedimentary sections. Unconformities occur throughout the sedimentary section, with major events in the late Oligocene and late Miocene which we correlate with suggested episodic uplift of the Transantarctic Mountains. Mesozoic and lower Paleogene rocks have not been sampled, but *Cooper et al.* [1987*a*] suspected the presence of these within their inferred deeper sedimentary section on the basis of recycled microfossils in drill cores, coastal glacial erratics, and seafloor cores in the eastern Ross Sea. However, on the basis of large offset seismic data discussed above, we speculate that these rocks may also comprise layered volcanic sequences. Basement rocks have been recovered only at DSDP site 270 (Figure 7*b*) on a basement high and are composed of early Paleozoic(?) gneiss [*Hayes and Davey*, 1975].

The uppermost part of the sedimentary section is cut regionally by the Ross Sea unconformity, which lies 2 to 42 m below the seafloor [*Karl et al.*, 1987] and spans the interval 14.7 to 4.0 Ma [*Savage and Ciesielski*, 1983]. Another major unconformity, which cuts the sedimentary sections and tops of basement ridges throughout the Ross Sea, lies at 3 to 6 km depth. *Hinz and Block* [1983] name this unconformity U6 in the Eastern basin and Central basin (Figure 17), and *Cooper et al.* [1987*a*] refer to it as the unconformity between acoustic units V4 and V5 in the Victoria Land basin. Reflection units below unconformity U6 are isolated within early rift stage grabens [*Cooper et al.*, 1991], whereas units above U6 cover the entire region. The early rift grabens probably extend south throughout the West Antarctic rift system.

Rocks have not been sampled from below the unconformity U6, with the exception of early Paleozoic(?) basement rock at DSDP site 270 (Figure 7*b*). Glacial sedimentary units have been sampled only above U6; similar units have been sampled only above basement highs. U6 is late Oligocene in age at DSDP site 270 on the Central high [*Hinz and Block,* 1983], but rocks from a unit above unconformity U6 are of early Oligocene age at core hole CIROS-1 (Figure 7*a*) on the uplifted flank of the Victoria Land basin [*Cooper et al.,* 1990]. Unconformity U6 therefore appears time-transgressive and is probably Eocene to Oligocene in age. Drilling data are needed to control the ages on these strata so that the timing on the stages of rifting and associated crustal extension can be determined.

Victoria Land Basin

The Victoria Land basin (Figures 7 and 17) is characterized by a thick (up to 14 km) layered section, presumed to be sedimentary by *Cooper et al.* [1987*a*] and considered by us to be partly volcanic. Extensive basement faulting (early rift) occurred in the Victoria Land basin and was followed by faulting and deformation of the sedimentary section in the Terror rift (late rift). In contrast, large offset (1 to 2 km) faults along the east flank of the early-rift graben do not disrupt the nearly flat-lying sedimentary strata that fill the graben (Figure 17), indicating that downfaulting of the basement occurred mostly prior to, rather than during, deposition of the overlying sedimentary/volcanic section. Strata above and below U6 are disrupted in the Terror rift by large normal faults that extend from the basement to the seafloor (Figure 17) [*Cooper et al.,* 1987*a*]. These faults and nearby subvolcanic intrusions (e.g., Figure 10) are thought to be late Paleogene and Neogene (e.g., late rift) features that were emplaced after deposition of most of the sedimentary section. Faulting along the west side of the Victoria Land basin has occurred in part during the Eocene and younger uplift of the Transantarctic Mountains and has resulted in at least three major angular unconformities [*Cooper et al.,* 1987*a*]. Their existence is further evidence for episodic uplift and extension.

Eastern and Central Basins

Seismic reflection data from the Eastern and Central basins (Figures 7 and 17) show that the maximum sedimentary thickness is 6–8 km [*Hinz and Block,* 1983]. Basement rocks beneath the center and flanks of these basins appear highly eroded on reflection interpretations. Deformation of the sedimentary section is generally minor throughout the basins, and reflectors are disrupted only locally within intrabasement grabens and over basement structures. The Central basin overlies extended crust, as is evident from large offset seismic data where the Moho is 17–19 km deep [*Trehu et al.,* 1989]. The short-wavelength magnetic anomalies correlated with shallow volcanic sources (Plate 2) that are present in the general area of the Central basin have not been crossed by seismic reflection profiles. Therefore the interpretive cartoon of Figure 17 does not indicate penetrative structures such as that labeled V in the Victoria Land basin.

Significance of Reflection Results to Study of West Antarctic Rift System

The seismic reflection results summarized in Figures 7 and 17 and discussed briefly here are relevant to the subject of the West Antarctic rift system specifically, as follows.

1. These data, tied to the sparse drill holes across the part of the rift underlying the Ross Sea shelf, provide the only age control for the ice-covered and sea-covered parts of the rift beneath the Ross Embayment and Byrd Subglacial Basin.
2. Rifting in the Ross Sea, and likely throughout the West Antarctic rift system, has been episodic. Most of the basement downfaulting and brittle crustal deformation occurred prior to development of unconformity U6 (Eocene to Oligocene time) and probably in late Mesozoic time [*Cooper et al.,* 1990, 1991].
3. The interpretation of horst and graben structures in the Ross Sea (Figure 17) probably extends beneath the Ross Ice Shelf and the Byrd Subglacial Basin.
4. The orientation of the structural basins underlying the Ross Sea shelf defined by seismic reflection results are parallel to the "grain" of the West Antarctic rift system (e.g., the Transantarctic Mountains shoulder, Plate 2, and Figures 2 and 9).
5. Together with the aeromagnetic data (Plate 2), the seismic reflection data provide the best available geophysical evidence for concealed submarine or subglacial late Cenozoic volcanic structures in the West Antarctic rift system.

DISCUSSION

The early reconnaissance geophysical studies of *Bentley* [1964], *Behrendt* [1964*a*], *Bentley and Clough* [1972], and *Bentley and Robertson* [1982] encompassed all of the interior of West Antarctica, and these data provide a provisional basis for inferring boundaries and basic characteristics of the entire rift system in West Antarctica. The gravity data suggesting crustal thinning beneath the West Antarctic rift (Byrd Subglacial Basin), which show similarities to those in the East African and Rio Grande rift systems, provide the basis for the suggestion that alkaline volcanism in Marie Byrd Land was related to the same major intracontinental rift system as the volcanoes in the western Ross Embayment [*LeMasurier*, 1978; *LeMasurier and Rex,* 1982, 1989].

Many of the recently published discussions of rifting cited in this paper [*Cooper et al.,* 1987*a*; *Stern and ten Brink,* 1989; *Tessensohn and Wörner*, 1991] have concentrated on specific areas (e.g., the Ross Embayment, western Ross Sea, and Transantarctic Mountains bordering it) and most of the data collected since 1950 and discussed here are also concentrated in the Ross Embayment area. However, we attempt to place our geophysical review in the larger context of the West Antarctic rift system using the earlier results. This world class rift system appears unusual compared with other rift systems in at least two respects: (1) even after allowing for isostatic adjustment after ice removal, the elevation is anomalously low (as is the North Sea rift system) throughout most of the area (Figure 2) compared to other rifts; and (2) Antarctica is nearly aseismic [*Kaminuma*, 1982; *Adams et al.*, 1985], whereas earthquakes are typically associated with active rifting. Although we cannot totally explain these observations, we offer the following suggestions.

Anomalous Low Elevations

In the rift model proposed by *Stern and ten Brink* [1989], they required a very low rigidity (4×10^{22} N m) for the lithosphere beneath the Ross Embayment, which is approximately equal in thickness to the crust (about 20 km) beneath the Ross Sea shelf (Figure 17), as interpreted from large-offset seismic studies. *Wörner and Viereck* [1989] in their discussions on Mount Melbourne (located in Figure 7 and Plate 1) noted evidence for relative and absolute subsidence of this volcano, in contrast to the high uplift rate of the Transantarctic Mountains. Although the Neogene sedimentary section in the Victoria Land basin is relatively thin [*Cooper et al.*, 1987*a*], the deep topographic depression of the continental shelf over the basin (Figure 17) suggests subsidence faster than sedimentary deposition and ice erosion. The high seismic velocity (7+ km/s) in both the upper and lower crust beneath the Ross Sea, as discussed above, can be interpreted as evidence for dense mafic underplating and crustal intrusion and would also contribute to rapid subsidence. All of these lines of evidence support a thin, hot, very weak, dense extended lithosphere which is subsiding rapidly. By comparison with the Bouguer anomaly map (Figure 13), we infer that similar crust and lithosphere is characteristic of the entire West Antarctic rift system and accounts for the anomalously low elevation. The Byrd Subglacial Basin area of the rift is filled with about 3 km of ice at about 1.5 km snow surface elevation. If the ice were compressed so that its density equaled that of low-density Neogene sedimentary rock, the surface elevation would be about sea level for the same mass load. This elevation should be compared with other rifts rather than the approximate mean 1.5 km depth of the base of the ice because the ice is isostatically compensated [*Bentley*, 1983]. Another area with consistently low elevation during active rifting is the Midcontinent rift system of North America [*Behrendt et al.*, 1988].

Low Seismicity

Although reasonably high quality seismographs have operated continuously since the International Geophysical Year (1956–1957), the only unambiguous earthquake, M_b 4.5 [*Adams et al.*, 1985], occurred in East Antarctica in 1982. *Bentley* [1983, 1991] discussed seismicity and rejected the argument of ice quakes for three >4 magnitude events next to major coastal outlet glaciers. He considers it more likely that these earthquakes were associated with fault zones that control the location of the glaciers. One of these earthquakes occurred in the Rennick graben (Figure 7), which *Roland and Tessensohn* [1987] interpreted as an early onshore branch of the rift system. We find it intuitively difficult to accept that stress buildup in ice alone could produce a >4 magnitude earthquake.

Johnston [1987] provided a reasonable explanation for the anomalously low seismicity of Antarctica and Greenland, on the basis of earthquake suppression by large continental ice sheets. *Johnston* [1987] proposed two theoretical explanations based on (1) the effect of static overburden (preferred) and (2) control of crustal pore pressure by insulation from a significant amount of meteoric water. The first explanation appears to require a compressive stress regime [*Johnston*, 1987], which is likely true for the continent as a whole because Antarctica is surrounded by spreading centers. However, the evidence for active extension associated with the West Antarctic rift system casts doubt on its applicability here. The second explanation, based on suppression of pore pressure by ice, would not seem to apply to the seawater-covered area of the Ross Embayment, although it might well apply in the thick ice-covered area of the Byrd Subglacial Basin.

Microseismicity is associated with active volcanoes such as Mount Erebus on Ross Island, and seismicity has been associated with a volcanic eruption on Deception Island [*Kaminuma*, 1982] at the north end of the Antarctic Peninsula (far from the West Antarctic rift system). Microearthquakes and "ultramicroearthquakes" of probable tectonic origins were reported to occur every 2 days [*Kaminuma*, 1982], measured by a station in the Transantarctic Mountains about 120 km from Mount Erebus (and thus too far to detect its activity). From stations on Ross Island *Rowe and Kienle* [1986] also reported tectonic microearthquakes which they associated with rifting. We associate these with faulting in the Victoria Land basin (Terror rift), as is suggested by the occurrence of Holocene fault scarps (such as shown in Figure 11). Greater frequency of microseismicity would probably be detected elsewhere within the West Antarctic rift system were more seismographs deployed. *Berg et al.* [1989] (from xenolith data) infer at present an anomalously high temperature at shallow depth. Possibly the very low rigidity of the extended crust of the Ross Embayment [*Stern and ten Brink*, 1989] and very high heat flow likely [*Berg et *al.*, 1989] in the area (including the Transantarctic Mountains) have led to rapid creep along faults, resulting in movement with few earthquakes. Therefore we do not consider the low seismicity significant evidence against the presence of an active rift; for example, the Rio Grande rift in the western United States would probably not be detectable solely from sparse seismograph coverage typical of that in Antarctica.

A fault with 300 m vertical offset cutting a Pliocene moraine in the Beardmore Glacier area [*McKelvey et al.*, 1991], numerous other faults cutting this moraine, and fault scarps cutting the seafloor (e.g., Figure 11) over the Victoria Land basin imply large earthquakes in the past, although one could consider aseismic creep similar to that reported for a fault scarp in Death Valley [*Sylvester and Bies*, 1986]. We infer that fault scarps, such as the one shown in Figure 11, must have formed after the latest deglaciation, or else they would have been eroded by overriding ice. *Muir-Wood* [1989] discussed evidence for earthquakes and rapid stress release shortly after deglaciation in Fennoscandia. He interpreted 5 to 6 orders of magnitude higher earthquake frequency followed by an anomalously seismically quiet period at present. Possibly *Muir-Wood*'s [1989] results for Fennoscandia provide some explanation for the fault scarps we see in reflection profiles and the absence of earthquakes now in the active West Antarctic rift system, i.e., high seismicity immediately following deglaciation and low seismicity at present.

Okal [1981] concluded that although the Antarctic continent is essentially aseismic, the Antarctic plate has three times the area of the continent alone and is not aseismic. The total seismic energy released from the plate during 1925–1980 is comparable to that of the African and Nazca plates, which have similar tectonics. Okal concluded that the ring of spreading ridges does transmit stress, which he intrepreted as clearly showing the importance of ridge push as a driving mechanism for plate movement. Ridge push would result in a horizontal compressive stress regime in Antarctica favoring *Johnston*'s [1987] static overburden hypothesis for lack of seismicity but does not explain the extension in the West Antarctic rift system. The stress situation (i.e., a ring of spreading ridges around Antarctica) is similar in the African plate, where the East African rift developed in spite of this fact. The West Antarctic rift system is probably an analog. We do not mean to imply that a ring of spreading centers caused either rift.

Zoback et al. [1989], in discussing global patterns of stress and constraints on intraplate deformation, noted that whereas asthenospheric drag is a major source of stress in a number of plates, ridge push and forces due to plate collision and subduction cannot be ruled out. Unfortunately, data for the Antarctic plate are too sparse to make any attempt to examine possible correlation of shear stress and plate velocity. For both the Antarctic and Africa plates absolute motion is very slow [*Okal*, 1981], so asthenospheric drag would likely be very low in both cases.

Uplift of Rift Shoulder

Active extension in the West Antarctic rift system at least since early Oligocene [*Barrett*, 1989] is implied by the volcanic activity (Figure 2) and uplift of the rift shoulder. Late Mesozoic–early Cenozoic extension and basin downfaulting were probably even greater than in late Cenozoic time [*Cooper et al.*, 1982; *Cooper et al.*, 1987*a*; *Fitzgerald et al.*, 1986]. The New Zealand–Campbell Plateau block broke away from Marie Byrd Land (about 80 Ma) [*Bradshaw*, 1989] when the major Ross Sea shelf grabens were likely downfaulted in the late Mesozoic. Uplift of the Transantarctic Mountains commenced about 60 Ma [*Fitzgerald*, 1989]. However, a growing body of evidence [*Behrendt and Cooper*, 1991] summarized in a previous section indicates rapid episodic uplift of the rift shoulder in the late Cenozoic and Holocene. We interpret this uplift along the Transantarctic Mountains bordering the Ross Embayment to mark the Cenozoic West Antarctic rift shoulder extending to the Ellsworth Mountains (but not to the Pensacola-Shackleton-Theron part of the Transantarctic Mountains) (Figure 3). If a maximum uplift of 10 km in northern Victoria Land [*Fitzgerald*, 1989] area is correct, then as much as 6 km of erosion has occurred in this area, resulting in the maximum altitude of about 4 km observed now. However, there is no direct evidence that the late Mesozoic and Cenozoic sedimentary rocks in basins beneath the adjacent continental shelf [*Cooper et al.*, 1987*a*] were ever deposited above the section exposed in the present rift shoulder ranges.

Mechanisms for Rifting

Because Antarctica is essentially surrounded by mid-ocean ridges, the only apparent source for stress to drive any tectonism is ridge push [*Okal*, 1981], yet it seems unlikely that this could be the source of the extension. Therefore another mechanism seems to be required. *Smith and Drewry* [1984] proposed that the rise of the Transantarctic Mountains is a delayed effect caused by overriding by East Antarctica of anomalously hot asthenosphere (hot spot) that formed under West Antarctica in Late Cretaceous time. In this model, the resultant increase in heat flow would result in phase changes in the upper mantle leading to uplift. *Fitzgerald et al.* [1986] pointed out that this model does not address the asymmetry (i.e., the high rift shoulder scarp on only one side) of the Transantarctic Mountains which they noted in the Ross Embayment.

In turn, *Fitzgerald et al.* [1986] proposed a simple shear model for uplift of the Transantarctic Mountains and extension observed in the basins beneath the Ross Sea shelf. They estimated about 200 km (25–30%) of extension, and their model required simple shear in the rift basins as well as along a master décollement extending from the Central high to beneath the Transantarctic Mountains. In their estimate [*Fitzgerald et al.*, 1986], they assumed an average of 25–30 km crustal thickness for the attenuated crust beneath the Ross Sea shelf. The large offset seismic results obtained over the Ross Sea shelf (Figure 15) suggest that the crust beneath the shelf may be only 17–21 km thick, requiring even greater extension than the 200 km proposed by *Fitzgerald et al.* [1986]. The Byrd Subglacial Basin–Ross Embayment was probably extended greatly during late Mesozoic time, but a significant part of the total could have occurred during Cenozoic time, and would fall within constraints from plate reconstruction models [*Stock,* 1989*a*; J. M. Stock, personal communication, 1989]. If, for example, a 40-km-thick crust were stretched to 20 km across the rift now presently approximately 750 km wide, during late Mesozoic-Cenozoic time, about 350–400 km total extension is implied. If a significant thickness of volcanic flows were extruded and underplating occurred, this would of course require still greater extension for a given present crustal thickness [e.g., *Behrendt et al.*, 1988, 1990] but would probably be localized (e.g., Figure 11) rather than uniformly distributed across the 750-km-wide rift system.

Stern and ten Brink [1989], although not disputing the asymmetry for the tectonic development of the Transantarctic Mountains, reject the simple shear model for faulting at the mountain front. They argue that this mechanism could not provide the observed magnitude (5 km) of uplift of the Transantarctic Mountains in the McMurdo area which they investigated. Instead they propose a model in which two lithospheric plates of vastly different effective thermal ages are juxtaposed and no shear stresses are transmitted across the boundary. They estimate flexural rigidities of 1×10^{25} and 4×10^{22} N m for the two plates, East Antarctica and the extended crust underlying the Ross Embayment, respectively. In the model these plates are separated at the mountain front by a stress-free boundary. Uplift of the rigid cantilevered beam is driven by heating at the free edge. Stern and ten Brink point out that their determinations are consistent with effective thermal ages for East Antarctica of about 600 Ma and the Ross Embayment of 25 Ma, which we find quite appropriate for the late Cenozoic volcanism observed.

Stern and ten Brink [1989] suggest that about 70 m.y. is required for an average density contrast of 1.5% associated with thermal conduction to penetrate 50 km horizontally beneath the Transantarctic Mountains, thus providing the necessary thermal uplift. They note that this is within the 90–65 Ma proposed by *Smith and Drewry* [1984] for the initiation of hot spot overriding. *Stern and ten Brink*'s [1989] proposed 70 m.y. also is consistent with *Fitzgerald*'s [1989] interpretation of the start of uplift of the Transantarctic Mountains at 60 Ma. They note that it could take several tens of millions of years of thermal conduction to weaken the edge of the East Antarctic plate and start the uplift process. The evidence for rapid uplift discussed above suggests that once it has started, it proceeds rapidly similar to other rift shoulders. This could account for the high Neogene uplift rates proposed here and by *Behrendt and Cooper* [1991]. The *Smith and Drewry* [1984] model has an additional attraction in that it suggests that the later volcanism observed in northern Victoria Land (and offshore according to the magnetic and seismic data discussed in this paper) would have started about 20 m.y. ago. They point out that this volcanism should be increasing in intensity, reaching its maximum development today, which is again consistent with the 25-Ma thermal age calculated for the Ross Embayment [*Stern and ten Brink,* 1989] and the rapid uplift discussed above. However, the early Oligocene (or older) date for Cenozoic volcanic rock recovered from the CIROS hole [*Barrett,* 1989] indicates that volcanism started at least by that time.

The *Stern and ten Brink* [1989] and *Smith and Drewry* [1984] discussions of heat flow are consistent with several of the observations from the geophysical data discussed in this paper, as follows: (1) evidence of extension seen throughout the area of the Ross Sea aeromagnetic survey (Figure 9 and Plate 2) and seismic surveys (Figure 17), (2) widespread submarine and subglacial volcanism inferred throughout the West Antarctic rift system, and (3) evidence of extension throughout the rift interpreted from the Bouguer anomaly data combined with the seismic depths to Moho beneath the Ross Sea shelf. Very high heat flow based on xenolith thermobarometry [*Berg et al.,* 1989], a 40°C/km gradient in the CIROS hole [*White,* 1989], and the sparse *Blackman et al.* [1987] heat flow data are consistent with the presence of shallow volcanism inferred from short-wavelength magnetic anomalies (Plates 1 and 2 and Figures 8 and 10) in the Victoria Land basin area. High temperature in the upper crust also is consistent with the negative magnetic anomaly over the Victoria Land basin, interpreted as possible evidence of thermal demagnetization [*Behrendt et al.,* 1987, 1991] of layered volcanic rocks which we infer to be present in the deepest 8 to 9 km of the basin. Possibly the extensive magmatic crustal intrusion and underplating which we interpret from the Bouguer anomalies result from a thermal plume mechanism [*LeMasurier and Rex,* 1989] such as that discussed by *White and McKenzie* [1989]. Marie Byrd Land (the Amundsen-Bellingshausen flank) uplift and subsidence seem closely related to magmatic activity. The plume mechanism proposed for this region rests on (1) the association of contemporaneous magmatism, centrifugal migration of felsic activity, and a stationary plate environment; (2) isotopic data which suggest an asthenospheric source; and (3) the fact that Marie Byrd Land basalts are geochemically indistinguishable from oceanic island basalts that are widely accepted as products of plume magmatism [*Futa and LeMasurier,* 1983; *LeMasurier and Rex,* 1989; *Hole and LeMasurier,* 1990].

Rifting in West Antarctica

We think it is likely that the arguments presented above for the Ross Embayment apply as well to the ice-covered areas of Marie Byrd Land and its exposed volcanic rocks and to the adjacent rift shoulder defined by the Transantarctic Mountains–Whitmore-Ellsworth Mountains escarpment. Unfortunately, the geophysical data are very sparse throughout this area.

There are two lines of reasoning which suggest that the Marie Byrd Land plume model may be extended to the late Cenozoic volcanic and tectonic activity of the entire rift system. First, *White and McKenzie* [1989] have shown that it is geophysically and geologically reasonable, in several rift localities in the world, to visualize a plume head having a diameter on the order of 2000 km. Second, *Hole and LeMasurier* [1990] have shown that there is a strong geochemical coherence to all the rift-related basalts in West Antarctica, from western Ellsworth Land through Marie Byrd Land to the western Ross Embayment. These basalts are all indistinguishable from plume-related oceanic island basalts but are clearly distinguishable from Antarctic Peninsula postsubduction basalts that are not related to the West Antarctic rift system.

The early-rift grabens of the Ross Sea (Figures 7 and 17) are separated by broad, eroded basement ridges; the area has probable crustal thicknesses of 17–21 km. One ridge (Central high) continues north and west into the south Pacific Ocean as Iselin Bank. Early-rift grabens of the Eastern and Central basins can be traced from near the continental shelf edge (~1000-m contour, illustrated in Figure 7) to the edge of the Ross Ice Shelf. These grabens can probably be correlated with similar structures beneath the Campbell Plateau [e.g., *Bradshaw*, 1989]. The Victoria Land basin, however, terminates at its north end against the Transantarctic Mountains (Figure 7).

The Polar 3 anomaly may indicate a transfer fault if a basin similar to the Victoria Land basin were present with a north trend extending north at the northeast end of the Polar 3 anomaly. The negative magnetic anomaly (Plate 2) suggests that there could also be a continuation of the Victoria Land basin in northern Victoria Land (for example, Rennick graben) [see *Roland and Tessensohn*, 1987]. Early-rift grabens beneath the Ross Sea continental shelf probably continue south beneath the Ross Ice Shelf and throughout the ice-covered area of the West Antarctic rift system [*LeMasurier and Rex*, 1983; *Jankowski et al.*, 1983; *Cooper et al.*, 1991]. *LeMasurier and Rex* [1983] pointed out that the ice in Marie Byrd Land–Byrd Subglacial Basin is in fact much of the "sedimentary rock" filling the basins there. The subice topography contours in Figure 2, shown in sufficient detail by *Drewry* [1983], illustrate that the seafloor contours are much smoother beneath the Ross Sea shelf and Ross Ice Shelf (probably because of the sediment filling of basins) than the subice topography of the Byrd Subglacial Basin underlying the moving ground ice there.

The 5 km of relief across the Transantarctic Mountains front in the northern Victoria Land–southern Victoria Land area continues from 4 to 7 km relief along the entire rift shoulder (Figures 2 and 3). This relief likely is related to the thermal structure extending throughout the West Antarctic rift system. Possibly the *Stern and ten Brink* [1989] model could be applied to the scarp marking the interpreted rift shoulder along the Horlick-Whitmore-Ellsworth Mountains trend. Whatever the mechanism, we interpret the high uplift of the rift shoulder and the Amundsen-Bellingshausen flank as resulting from lateral thermal heating from the upwarped shallow asthenosphere beneath the Byrd Subglacial Basin and Ross Embayment.

In our discussion above of gravity and the sparse seismic refraction data in the West Antarctic rift system, we inferred that shallow- to lower-crustal intrusions are the sources of steep gravity gradients and high seismic velocities. If underplating and lower-crustal intrusions are present, they may account for a few deep reflections from beneath McMurdo Sound and the Ross Sea shelf [*McGinnis et al.*, 1985; *Kim et al.*, 1986; *Cooper et al.*, 1987*b*; *Trehu et al.*, 1989] as has been interpreted for rift systems elsewhere than Antarctica [e.g., *Behrendt et al.*, 1988, 1990]. The measured velocities interpreted as evidence for shallow- and lower-crustal intrusions (and/or underplating) range from about 6.7 to 7.4 km/s, as discussed in the seismic section. *Berg et al.* [1989] reported gabbro norite cumulates from xenoliths collected from volcanic rocks in the McMurdo area, and *Wörner and Viereck* [1989] reported similar rocks with xenoliths near Mount Melbourne. Both finds support the presence of mafic intrusions. The Polar 3 and other >1000-nT magnetic anomalies (Figure 2) are evidence for mafic (or ultramafic?) magmatic rocks. The only measurement of *Pn* (Moho) is 8.2 km/s in the McMurdo Sound area [*McGinnis et al.*, 1985], a velocity which seems high for an area of active rifting.

The late Cenozoic alkaline volcanic rocks exposed at sparse locations (Figure 2) along the West Antarctic rift system and in the CIROS core hole provide the only age control for Cenozoic magmatism (early Oligocene or earlier to present) associated with rifting. The magmas probably resulted from decompression melting of upwarped asthenosphere beneath the attenuated rifted crust [*Futa and LeMasurier*, 1983; *LeMasurier and Rex*, 1989; *Schmidt-Thomé et al.*, 1990; *Mueller et al.*, 1990]. On the basis of magnetic data, we interpret these volcanic rocks to occur over a much broader area (Figures 2, 4, and 5 and Plate 1) of the ice-covered West Antarctic rift system than their areas of outcrop. We interpret the most extensive area of rift-related volcanism beneath the ice-covered area of Marie Byrd Land to be that marked by the abundance of short-wavelength anomalies (shallow source) discussed in the magnetics section (Figure 5).

We realize that extending the tectonic interpretation for Cenozoic rifting and crustal thinning developed for the Ross Sea–northern Victoria Land area throughout the West Antarctic rift system is speculative. We hope, however, that these ideas may stimulate acquisition of badly needed geophysical data and encourage others to interpret their work throughout the area within the context of this model.

Progression of Rifting From Jurassic to Present

LeMasurier [1990*a*, *b*] noted above that the age of the oldest Cenozoic volcanism is about early Oligocene (or earlier) in the Ross Embayment and progressively decreases toward the end of the rift near the Antarctic Peninsula where it is Pliocene in age. The age of earliest rifting in the Ross Sea basins is not known but is probably late Mesozoic. We suggest that the Gondwana breakup and the West Antarctic rift system are part of a continuously operating single system. There is a progression, as also noted by *Lawver et al.* [1991], in rifting and separation around East Antarctica from Jurassic (179–162 Ma) for Africa [e.g., *Ford and Kistler,* 1980; *White and McKenzie,* 1989] to Cretaceous (about 130 Ma) for Greater India [*Johnson et al.,* 1976] and about 110–90 Ma for Australia [*Cande and Mutter,* 1982] to Late Cretaceous–early Cenozoic (about 95 to 85 Ma) [*Bradshaw,* 1989] for New Zealand and the Campbell Plateau from Marie Byrd Land. We propose that this rifting has propagated like the peeling of an orange, extending into West Antarctica (with a spreading center jump) to its present location in the Ross Embayment [*Davey,* 1981] and West Antarctica. This idea is generally supported by the structures between the Ross Sea and New Zealand revealed in the Geosat data [*Sandwell and McAdoo,* 1988] and the suggestions of thermal activity beginning in West Antarctica in Late Cretaceous time [*Smith and Drewry,* 1984].

Most extension in the rift in West Antarctica is probably late Mesozoic in age, but a significant part of the total could have occurred during the Cenozoic. Much of the rift process, including all the dated alkaline volcanism, took place after the separation of Antarctica from Australia and New Zealand. As it is a process restricted to the Antarctic plate with no apparent link to the surrounding plate tectonic features, the cause for the rifting will have to be assumed to lie in mantle processes under the stationary Antarctic plate.

Additional Studies

Specifically, some of the additional studies necessary in the West Antarctic rift area are as follows:

1. Aeromagnetic surveys similar to the survey shown in Plate 1 combined with radar ice-sounding data over grounded ice, along transect zones crossing the rift.
2. Gravity traverses (airborne or surface) with ice thickness determinations over ice-covered areas and on outcrops across the steep Bouguer anomaly gradient (as shown in Figure 13) over the rift shoulder adjacent to the Horlick-Whitmore-Ellsworth Mountains area to determine if the 4–7 mGal/km gradient present along parts of the Ross Embayment is continuous to this area. This is important to help resolve the ambiguity whether the source of the steep gradient is associated with early Jurassic or Cenozoic rifting.
3. Large offset seismic surveys over the rift shoulder and extended crust in the thick ice area of Marie Byrd Land (Byrd Subglacial Basin and Ross Embayment) to measure crustal thickness and determine the areal extent of the ~7-km/s refractors.
4. Deep crustal (20-s recording time) multichannel (vertical incidence) reflection profiles in the Ross Embayment–Marie Byrd Land area. A deep profile could be done economically using marine techniques over the Ross Sea shelf but would be much more expensive in the ice-covered areas.
5. Heat flow measurements beneath the continental shelf and the floating and grounded ice sheet. If this is not possible, electrical measurements could provide indirect thermal structure determination.
6. In addition to the geophysical surveys, drill holes are needed in the basin and platform areas beneath the Ross Sea shelf (Figure 7), particularly to put constraints on the ages of the interpreted unconformities (Figure 17). This information is needed to better define the timing of the West Antarctic rift system, including the early rift stages (which are probably as old as late Mesozoic). If ever feasible, drilling is needed through the ice filling the Byrd Subglacial Basin and into the rock (including rift basins) beneath with shallow or preferably deep penetration.

SUMMARY

In this paper, we have attempted to review all of the geophysical results and better define the little known West Antarctic rift system, one of the great rift systems of the world. The major characteristics are as follows:

1. The West Antarctic rift system covers an area about 3000 by 750 km through West Antarctica. It is comparable in size to the Basin and Range in North America or the East African rift system, as pointed out by *LeMasurier* [1978, 1990*a*] and *Tessensohn and Wörner* [1991].
2. Bimodal alkaline volcanic rocks typical of other rifts, erupted from the Oligocene (or earlier) to the present, crop out throughout the West Antarctic rift system, and are probably widely present beneath water- and ice-covered areas. Late Cenozoic magmatic rocks are exposed (1) along the front of the Transantarctic Mountains from northern Victoria Land, being geographically coincident with Jurassic tholeiites (Ferrar dolerites and Kirkpatrick basalts), to about the Horlick Mountains but do not exist any further along the Transantarctic Mountains and (2) throughout the mountains of Marie Byrd Land to the southern Antarctic Peninsula. Magnetic data suggest the presence of Cenozoic volcanic rocks beneath the western part of the Byrd Subglacial Basin and indicate their absence beneath the eastern part of the basin (closest to the Ellsworth-Whitmore Mountains).

Magnetic data also suggest Cenozoic volcanic rocks are present on the Ross Sea shelf and possibly beneath the Ross Ice Shelf.

3. The rift system is marked by a spectacular physiographic scarp coincident with the Transantarctic Mountains in Victoria Land (about 5 km relief) through to the Ellsworth-Whitmore-Horlick Mountains where relief is as great as 7 km. The Amundsen-Bellingshausen flank of the rift system is lower and more internally fragmented by horst and graben structures, while the axial part or "rift valley" comprising early rift grabens is buried beneath the Ross Sea continental shelf, the Ross Ice Shelf, and the West Antarctic ice sheet. Rifting probably started in the late Mesozoic and has continued through the Cenozoic. However, the oldest known associated magmatic rocks are only about Oligocene or earlier.

4. There is a distinct difference between the elevations of mountains (rift shoulder) flanking the West Antarctic rift which were uplifted during late Cenozoic extension and elevations of those ranges in the Transantarctic Mountains beyond the Horlick Mountains toward the Weddell Sea not affected by late Cenozoic extension. The maximum elevations along the rift shoulder are about 4 km in Victoria Land, 3 km in the Horlick-Whitmore Mountains, and 5 km in the Ellsworth Mountains. In contrast, the maximum elevation in the Thiel and Pensacola Mountains is 2 km decreasing toward the Weddell Sea to $1\frac{1}{2}$ and 1 km in the Shackleton and Theron ranges, respectively. Uplift and faulting in this low area of the Transantarctic Mountains probably is substantially older than that which occurred along the highest parts of the Cenozoic West Antarctic rift shoulder.

5. Although not definitive, a growing body of physiographic, geophysical, glacial geomorphological, and paleontological evidence from the Transantarctic Mountains along the Ross Embayment leads us to the conclusion that the rift shoulder has been rising since about 60 Ma at episodic uplift rates of the order of 1 km/m.y. most recently since the mid-Pliocene rather than continuously at the mean rate of 100 m/m.y. This uplift rate could account for the present 4–5 km highest elevation since 2–5 Ma, but evidence also indicates earlier episodes of uplift and erosion. We suggest a possible climate forcing effect on the advance of the Antarctic ice sheet.

6. Although high-amplitude (>100 nT) short-wavelength magnetic anomalies are caused by both the late Cenozoic alkaline volcanic rocks and the Jurassic volcanic rocks (but not the Ferrar dolerite sills), magnetic data are useful in places in approximately delineating the extent of Cenozoic magmatism beneath the ice. Over the Ross Sea continental shelf, a detailed aeromagnetic survey reveals the presence of about 100 inferred shallow-source submarine volcanic edifices (or subvolcanic intrusions) along linear trends interpreted as rift fabric. Conspicuous shallow-source magnetic anomalies on widely spaced profiles over the ice-covered areas of Marie Byrd Land (correlated with late Cenozoic volcanoes in outcrop areas) suggest the presence of Cenozoic volcanism throughout a large area there. The absence of such shallow-source anomalies east of a prominent north trending break in the high-amplitude anomaly pattern over the Byrd Subglacial Basin indicates the absence of Cenozoic volcanism to the east of this line (i.e., toward the Ellsworth Mountains). The smooth magnetic field is probably caused by the thick nonmagnetic (lower Paleozoic) metasedimentary rock cropping out in the Ellsworth Mountains.

Several linear (several hundred kilometers long) 1000+ nT anomalies (e.g., the Polar 3 anomaly at the front of the Transantarctic Mountains in Victoria Land) exist throughout the West Antarctic rift. We suggest this is evidence for possible oblique slip within this extensional system. We speculate that the "Sinuous Ridge anomaly" in the deepest part of the Byrd Subglacial Basin represents a change from alkaline to tholeiitic magmatism and the start of formation of an oceanic spreading center.

7. One of the long-recognized great gravity anomalies of the world marks the transition from West to East Antarctica across the rift shoulder in the Ross Embayment part of the West Antarctic rift system. Compared with East Antarctica, Bouguer anomalies increase by as much as 200 mGal over the attenuated rifted crust underlying the Ross Sea shelf, Ross Ice Shelf, and ice-covered part of Marie Byrd Land (Byrd Subglacial Basin). Although data are sparse, it is apparent that this anomaly continues along the Ellsworth-Whitmore part of the rift shoulder. In contrast, the difference in Bouguer gravity across the West Antarctica–East Antarctica boundary is lower (only about 130 mGal maximum range) in the Pensacola Mountains–Filchner Ice Shelf area outside the Cenozoic rift area.

Although primarily the result of crustal thinning through rifting beneath the Ross Embayment, the steep gradients (4–7 mGal/km) measured at a few places (Beardmore Glacier, McMurdo, and northern Victoria Land areas) require high-density rock within the rifted crust, which we suggest is evidence of shallow- or lower-crustal intrusion of mafic or ultramafic rock associated with rifting. In contrast, the gravity gradient across the front of the Transantarctic Mountains in the Pensacola Mountains is only about 2 mGal/km, which is easily explained by a 15°–20° dip on the Moho. The thinned crust on the West Antarctic side is probably due to rifting, but whether this occurred in the Jurassic (post–Dufek intrusion) or later is unknown.

8. At several places close to the Horlick Mountains and the Whitmore Mountains part of the rift shoulder, and in the Ross Embayment–Ross Sea area, sparse seismic refraction data have shown evidence of 6.8–7.3 km/s velocity layers at relatively shallow depth within the crust. We consider this supportive of the gravity interpretation of high-density mafic (or ultramafic?) intrusions within the rifted crust. We cannot say whether these inferred intrusions resulted from Cenozoic or Mesozoic magmatic activity, but

they are likely to have been emplaced during major stages of extension, crustal thinning, and graben formation. The presence of similar high (6.9–7.4 km/s) velocities in the lower crust beneath the extended Ross Sea shelf suggests mafic (or ultramafic?) underplating there and likely elsewhere beneath the rift in ice-covered areas. In the 14-km-deep Victoria Land basin, high seismic velocities at shallow depths are interpreted as evidence of layered volcanic rather than sedimentary rocks, consistent with evidence of magmatic activity associated with rifting.

9. Reflections from Moho at several locations over the Ross Sea shelf and one at the northern Victoria Land mountain front show Moho depths from 17 to 22 km, consistent with our interpretation of crustal thinning caused by rifting. An uppermost mantle velocity of 8.2 km/s in McMurdo Sound is the only determination in the West Antarctic rift system area. The generally constant level in the Bouguer gravity map from the Ross Sea shelf through the Ross Embayment–Byrd Subglacial Basin to the Bellingshausen Sea indicates that the ~20-km-thick rift stage crust extends through this area as well and is consistent with the presence of early-rift grabens (see characteristic 3 above).

10. A combination of active volcanism, rapid topographic uplift at the rift shoulder, xenolith thermobarometry, interpretation of possible upwarp of Curie isotherm beneath the Victoria Land basin, a few heat flow measurements, and a high geothermal gradient (40°C/km, *White,* 1989) are all indicative of high heat flow in the Ross Embayment, characteristic of other young rift systems. The evidence of high temperatures at shallow depth beneath the Ross Sea continental shelf and adjacent Transantarctic Mountains is supportive of the model of thermal uplift associated with lateral heat conduction from the rift [*Stern and ten Brink,* 1989]. Possibly in combination with hot spot–thermal plume models this model can also explain the volcanism, rifting, and high elevation of the rift shoulder (Ellsworth Mountains) in greater Marie Byrd Land (including the Byrd Subglacial Basin).

We propose that the active West Antarctic rift system is a continuation of rifting that began in the Jurassic at the Weddell Sea end of the Transantarctic Mountains with separation of Antarctica from Africa and has continued clockwise around the present continental margin (with a spreading center jump) to its present location in the Ross Embayment and the Byrd Subglacial Basin. The rift system is active now and may have been so since the late Mesozoic.

Most extension in the rift in West Antarctica is probably late Mesozoic in age but a significant part of the total could have occurred during the Cenozoic. If, as an extreme case, a 40-km-thick crust thinned to 20 km across the present 750-km-wide rift during late Mesozoic-Cenozoic time, a maximum of about 350-400 km total extension is implied. Significant underplating would require still greater extension. The major part of the rift process, including all the dated alkaline volcanism, took place after the separation of Antarctica from Australia and New Zealand. As it is a process restricted to the Antarctic plate with no apparent link to the surrounding plate tectonic features, the cause for the rifting must lie in mantle processes under the stationary Antarctic plate.

Acknowledgments. We thank Charles Bentley for assistance in obtaining some of the early gravity data (from the 1950s and 1960s). Baerbel Lucchitta of the U.S. Geological Survey (USGS), Flagstaff, Arizona, provided the aeromagnetic-enhanced Landsat images used. Discussions with our colleagues—particularly Peter Barker, Peter Barrett, Charles Bentley, David Harwood, Karl Kellogg, Philip Kyle, Lisa Morgan, Dwight Schmidt, Hans Schouten, Brian Storey, Gerhard Wörner, and Mary Lou Zoback—were helpful. David Harwood, P. Mueller, M. Schmidt-Thomé, and Cynthia Ebinger furnished unpublished manuscripts. Karl Kellogg, John Grow, Larry Lawver, Kevin Furlong, and others reviewed the manuscript of this paper. The National Science Foundation (NSF) and the Alfred Wegner Institute in Germany provided logistic support for the Bundesanstalt für Geowissenschaften und Rohstoffe–USGS Antarctic surveys. W.E.L. gratefully acknowledges the support of NSF through grant DPP 8020836 and the hospitality of the British Antarctic Survey during the preparation of this paper.

REFERENCES

Adams, R. D., A. A. Hughes, and B. M. Zhang, A confirmed earthquake in continental Antarctica, *Geophys. J. R. Astron. Soc., 81,* 489–492, 1985.

Barrett, P. J., Antarctic Cenozoic history from the MSSTS-1 drillhole, western McMurdo Sound, *N. Z. DSIR Bull., 237,* 174, 1986.

Barrett, P. J. (Ed.), Antarctic Cenozoic history from the CIROS-1 drillhole, McMurdo Sound, *N. Z. DSIR Bull., 245,* 254, 1989.

Barrett, P. J., D. P. Elston, D. M. Harwood, B. C. McKelvey, and P. N. Webb, Mid-Cenozoic record of glaciation and sea-level change on the margin of the Victoria Land basin, Antarctica, *Geology, 15,* 634–637, 1987.

Barrett, P. J., M. J. Hambrey, and P. R. Robinson, Cenozoic glacial and tectonic history from CIROS-1, McMurdo Sound, in *Geological Evolution of Antarctica,* edited by M. R. A. Thomson, J. A. Crane, and J. W. Thomson, pp. 651–656, Cambridge University Press, New York, 1991.

Behrendt, J. C., Distribution of narrow-width magnetic anomalies in Antarctica, *Science, 144,* 995–999, 1964*a*.

Behrendt, J. C., Crustal geology of Ellsworth Land and the southern Antarctic Peninsula from gravity and magnetic anomalies, *J. Geophys. Res., 69,* 2047–2063, 1964*b*.

Behrendt, J. C., Magnetic anomalies of short wavelength, in *Magnetic and Gravity Maps of Antarctica, Antarct. Map Folio Ser.,* folio 9, edited by J. C. Behrendt and C. R. Bentley, plate 7, American Geographical Society, New York, 1968.

Behrendt, J. C., and C. R. Bentley (Eds.), *Magnetic and Gravity Maps of Antarctica, Antarct. Map Folio Ser.,* folio 9, American Geographical Society, New York, 1968.

Behrendt, J. C., and A. Cooper, Evidence of rapid Cenozoic uplift of the shoulder escarpment of the Cenozoic West Antarctic rift system and a speculation on possible climate forcing, *Geology, 19,* 315–319, 1991.

Behrendt, J. C., and R. J. Wold, Depth to magnetic 'basement' in West Antarctica, *J. Geophys. Res., 68,* 1145–1153, 1963.

Behrendt, J. C., J. R. Henderson, L. J. Meister, and W. Rambo, Geophysical investigations of the Pensacola Mountains and adjacent glacierized areas of Antarctica, *U.S. Geol. Surv. Prof. Pap., 844,* 27 pp., 1974.

Behrendt, J. C., D. J. Drewry, E. Jankowski, and M. S. Grim, Aeromagnetic and radio echo ice-sounding measurements show much greater area of the Dufek intrusion, Antarctica, *Science, 209,* 1014–1017, 1980.

Behrendt, J. C., D. J. Drewry, E. Jankowski, and M. S. Grim, Aeromagnetic and radio echo ice-sounding measurements over the Dufek intrusion, Antarctica, *J. Geophys. Res., 86,* 3014–3020, 1981.

Behrendt, J. C., A. K. Cooper, and A. Yuan, Interpretation of marine magnetic gradiometer and multichannel seismic-reflection observations over the western Ross Sea shelf, Antarctica, in *The Antarctic Continental Margin: Geology and Geophysics of the Western Ross Sea, Earth Sci. Ser.,* vol. 5B, edited by A. K. Cooper and F. J. Davey, pp. 155–178, Circum-Pacific Council for Energy and Natural Resources, Houston, Tex., 1987.

Behrendt, J. C., A. G. Green, W. F. Cannon, D. R. Hutchinson, M. W. Lee, B. Milkereit, W. F. Agena, and C. Spencer, Crustal structure of the Midcontinent Rift system: Results from GLIMPCE deep seismic reflection profiles, *Geology, 16,* 81–85, 1988.

Behrendt, J. C., F. Tessenhohn, W. LeMasurier, A. Trehu, and A. Cooper, Geophysical investigations of the Cenozoic West Antarctic rift system (abstract), *Eos Trans. AGU, 70,* 1344, 1989.

Behrendt, J. C., D. R. Hutchinson, M. Lee, A. Trehu, C. R. Thornber, W. Cannon, and A. G. Green, GLIMPCE seismic evidence of deep crustal and upper mantle intrusions and magmatic underplating associated with the Midcontinent Rift system of North America, *Tectonophysics, 173,* 595–615, 1990.

Behrendt, J. C., H. J. Duerbaum, D. Damaske, R. Saltus, W. Bosum, and A. K. Cooper, Extensive volcanism and related tectonism beneath the western Ross Sea continental shelf, Antarctic, Interpretation of an aeromagnetic survey, in *Geological Evolution of Antarctica,* edited by M. R. A. Thomson, J. A. Crane, and J. W. Thomson, pp. 299–304, Cambridge University Press, New York, 1991.

Bentley, C. R., The structure of Antarctica and its cover, in *Research in Geophysics,* vol. 2, *Solid Earth and Interface Phenomena,* edited by H. Odishaw, pp. 375–389, MIT Press, Cambridge, Mass., 1964.

Bentley, C. R., Free-air gravity anomalies, in *Magnetic and Gravity Maps of Antarctica, Antarct. Map Folio Ser.,* folio 9, edited by J. C. Behrendt and C. R. Bentley, plate 8, American Geographical Society, New York, 1968.

Bentley, C. R., Crustal structure of Antarctica, *Tectonophysics, 20,* 229–240, 1973.

Bentley, C. R., Crustal structure of Antarctica from geophysical evidence—A review, in *Antarctic Earth Science,* edited by R. L. Oliver, P. R. James, and J. B. Jago, pp. 491–497, Australian Academy of Science, Canberra, 1983.

Bentley, C. R., Configuration and structure of the subglacial crust, in *Antarctic Geology,* edited by R. J. Tingey, Oxford University Press, New York, in press, 1991.

Bentley, C. R., and J. W. Clough, Seismic refraction shooting in Ellsworth and Dronning Maud Land, in *Antarctic Geology and Geophysics, Int. Univ. of Geol. Sci. Ser. B,* vol. 1, edited by R. J. Adie, pp. 683–691, Universitetsforlaget, Oslo, 1972.

Bentley, C. R., and N. A. Ostenso, Glacial and subglacial topography of West Antarctica, *J. Glaciol., 3,* 882–911, 1961.

Bentley, C. R., and J. D. Robertson, Isostatic gravity anomalies in West Antarctica, in *Antarctic Geoscience,* edited by C. Craddock, pp. 949–954, University of Wisconsin Press, Madison, 1982.

Bentley, C. R., A. P. Crary, E. Thiel, and N. A. Ostenso, Structure of West Antarctica, *Science, 131,* 131–136, 1960.

Berg, J. H., and D. L. Herz, Thermobarometry of two-pyroxene granulite inclusions in Cenozoic volcanic rocks of the McMurdo Sound region, *Antarct. J. U.S., 21,* 19–20, 1986.

Berg, J. H., R. J. Moscati, and D. L. Herz, A petrologic geotherm from a continental rift in Antarctica, *Earth Planet. Sci. Lett., 93,* 98–108, 1989.

BGR-USGS, Total magnetic anomaly maps, Victoria Land–Ross Sea, Antarctica, scale 1:250,000, 7 sheets, U.S. Geol. Surv., Reston, Va., 1987.

Blackman, D. K., R. P. Von Herzen, and L. A. Lawver, Heat flow and tectonics in the western Ross Sea, Antarctica, in *The Antarctic Continental Margin: Geology and Geophysics of the Western Ross Sea, Earth Sci. Ser.,* vol. 5B, edited by A. K. Cooper and F. J. Davey, pp. 179–189, Circum-Pacific Council for Energy and Natural Resources, Houston, Tex., 1987.

Blakely, R. J., and R. W. Simpson, Locating edges of source bodies from magnetic or gravity anomalies, *Geophysics, 51,* 1494–1498, 1986.

Bosum, W., D. Damaske, N. W. Roland, J. C. Behrendt, and R. Saltus, The GANOVEX IV Victoria Land/Ross Sea aeromagnetic survey: Interpretation of anomalies, *Geol. Jahrb., Reihe E, 38,* 153–230, 1989.

Bradshaw, J. D., Cretaceous geotectonic patterns in the New Zealand region, *Tectonics, 8,* 803–820, 1989.

Cande, S. C., and J. C. Mutter, A revised identification of the oldest sea-floor spreading anomalies between Australia and Antarctica, *Earth Planet. Sci. Lett., 58,* 151–160, 1982.

Cooper, A. K., F. J. Davey, and J. C. Behrendt, Seismic stratigraphy and structure of the Victoria Land basin, western Ross Sea, Antarctica, in *The Antarctic Continental Margin: Geology and Geophysics of the Western Ross Sea, Earth Sci. Ser.,* vol. 5B, edited by A. K. Cooper and F. J. Davey, pp. 27–76, Circum-Pacific Council for Energy and Natural Resources, Houston, Tex., 1987*a*.

Cooper, A. K., F. J. Davey, and G. R. Cochrane, Structure of extensionally rifted crust beneath the western Ross Sea and Iselin Bank, Antarctica, from sonobuoy seismic data, in *The Antarctic Continental Margin: Geology and Geophysics of the Western Ross Sea, Earth Sci. Ser.,* vol. 5B, edited by A. K. Cooper and F. J. Davey, pp. 93–118, Circum-Pacific Council for Energy and Natural Resources, Houston, Tex., 1987*b*.

Cooper, A. K., F. J. Davey, and K. Hinz, Geology and hydrocarbon potential of the Ross Sea, Antarctica, in *Proceedings of the AAPG Antarctic Symposium,* edited by B. St. John, pp. 47–68, American Association of Petroleum Geologists, Tulsa, Okla., 1990.

Cooper, A. K., F. J. Davey, and K. Hinz, Crustal extension and origin of sedimentary basins beneath the Ross Sea and Ross Ice shelf, Antarctica, in *Geological Evolution of Antarctica,* edited by M. R. A. Thomson, J. A. Crane, and J. W. Thomson, pp. 285–292, Cambridge University Press, New York, 1991.

Cooper, R. A., C. A. Landis, W. E. LeMasurier, and I. G. Speden, Geologic history and regional patterns in New Zealand and West Antarctica—Their paleotectonic and paleogeographic significance, in *Antarctic Geoscience,* edited by C. Craddock, pp. 43–53, University of Wisconsin Press, Madison, 1982.

Cordell, L., and V. J. S. Grauch, Mapping basement magnetization zones from aeromagnetic data in the San Juan Basin, New Mexico, in *The Utility of Regional Gravity and Magnetic Anomaly Maps,* edited by W. J. Hinze, pp. 181–197, Society of Exploration Geophysicists, Tulsa, Okla., 1985.

Craddock, C., T. W. Bastien, and R. H. Rutford, Geology of the Jones Mountains area, in *Antarctic Geology,* edited by R. J. Adie, pp. 171–187, North Holland, Amsterdam, 1964.

Craddock, C., et al., *Geologic Maps of Antarctica, 1969–1970, Antarct. Map Folio Ser.,* folio 12, edited by V. C. Bushnell and C. Craddock, American Geographical Society, New York, 1969.

Dalziel, I. W. D., S. W. Garrett, A. M. Grunow, R. J. Pankhurst, B. C. Storey, and W. R. Vennum, The Ellsworth-Whitmore Mountains crustal block: Its role in the tectonic evolution of West Antarctica, in *Gondwana Six: Structure, Tectonics, and Geophysics, Geophys. Monogr. Ser.,* vol. 40, edited by G. D. McKenzie, pp. 173–182, AGU, Washington, D. C., 1987.

Davey, F. J., Geophysical studies in the Ross Sea region, *J. R. Soc. N. Z., 11,* 465–479, 1981.

Davey, F. J., Geology and structure of the Ross Sea region, in *The Antarctic Continental Margin: Geology and Geophysics of the Western Ross Sea, Earth Sci. Ser.,* vol. 5B, edited by A. K. Cooper and F. J. Davey, pp. 1–16, Circum-Pacific Council for Energy and Natural Resources, Houston, Tex., 1987.

Davey, F. J., and A. K. Cooper, Gravity studies of the Victoria Land Basin and Iselin Bank, in *The Antarctic Continental Margin: Geology and Geophysics of the Western Ross Sea, Earth Sci. Ser.,* vol. 5B, edited by A. K. Cooper and F. J. Davey, pp. 119–138, Circum-Pacific Council for Energy and Natural Resources, Houston, Tex., 1987.

Davey, F. J., D. J. Bennett, and R. E. Houtz, Sedimentary basins of the Ross Sea, Antarctica, *N. Z. J. Geol. Geophys., 25,* 245–255, 1982.

Davey, F. J., K. Hinz, and H. Schroeder, Sedimentary basins of the Ross Sea, Antarctica, in *Antarctic Earth Science,* edited by R. L. Oliver, P. R. James, and J. B. Jago, pp. 533–538, Australian Academy of Science, Canberra, 1983.

Debenham, F., Geological history of south Victoria Land, in *Scott's Last Expedition,* vol. 2, pp. 430–437, Smith Elder, London, 1913.

Denton, G. H., M. L. Prentice, D. E. Kellogg, and T. B. Kellogg, Late Tertiary history of the Antarctic ice sheet: Evidence from the dry valleys, *Geology, 12,* 263–267, 1984.

Doake, C. S. M., R. D. Crabtree, and I. W. D. Dalziel, Subglacial morphology between Ellsworth Mountains and Antarctic Peninsula: New data and tectonic significance, in *Antarctic Earth Science,* edited by R. L. Oliver, P. R. James, and J. B. Jago, pp. 270–273, Australian Academy of Science, Canberra, 1983.

Drewry, D. J., Antarctica: Glaciological and geophysical folio, 9 sheets, Scott Polar Res. Inst., Univ. of Cambridge, Cambridge, England, 1983.

Duerbaum, H. J., G. Druivenga, H. Geipel, and G. Merkel, Gravity measurements along a traverse from Mt. Melbourne to the Polar Plateau in north Victoria Land, Antarctica, *Geol. Jahrb., Reihe E, 38,* 231–243, 1989.

Ebinger, C. J., Geometric and kinematic development of border faults and accommodation zones, Kivu-Rusizi rift, Africa, *Tectonics, 8,* 117–133, 1989.

England, P., and G. Houseman, Extension during continental convergence, with application to the Tibetan plateau, *J. Geophys. Res., 94,* 17,561–17,579, 1989.

Fitzgerald, P. G., Uplift and formation of Transantarctic Mountains: Application of apatite fission track analysis to tectonic problems, paper presented at 28th International Geological Congress, Washington, D. C., July 9–19, 1989.

Fitzgerald, P. G., M. Sandiford, P. J. Barrett, and A. J. W. Gleadow, Asymmetric extensions associated with uplift and subsidence in the Transantarctic Mountains and the Ross Embayment, *Earth Planet. Sci. Lett., 81,* 67–78, 1986.

Ford, A. B., Fit of Gondwana continents—Drift reconstruction from the Antarctic continental viewpoint, *Int. Geol. Congr., 24th Proc. Sect. 3,* 113–121, 1972.

Ford, A. B., and R. W. Kistler, K-Ar age, composition, and origin of Mesozoic mafic rocks related to Ferrar Group, Pensacola Mountains, Antarctic, *N. Z. J. Geol. Geophys., 23,* 371–390, 1980.

Furlong, K. P., and D. M. Fountain, Continental crustal underplating: Thermal considerations and seismic-petrologic consequences, *J. Geophys. Res., 91,* 8285–8294, 1986.

Futa, K., and W. E. LeMasurier, Nd and Sr isotopic studies on Cenozoic mafic lavas from West Antarctica: Another source for continental alkali basalts, *Contrib. Mineral. Petrol., 83,* 38–44, 1983.

Garrett, S. W., L. D. B. Herrod, and D. R. Mantripp, Crustal structure of the area around Haag Nunataks, West Antarctica: New aeromagnetic and bedrock elevation data, in *Gondwana Six: Structure, Tectonics, and Geophysics, Geophys. Monogr. Ser.,* vol. 40, edited by G. D. McKenzie, pp. 109–115, AGU, Washington, D. C., 1987.

Garrett, S. W., M. P. Mashanyj, and D. Damaske, Interpretation of aeromagnetic data from the Ellsworth Mountains–Thiel Mountains ridge, West Antarctica, *J. Geol. Soc. London, 145,* 1009–1017, 1988.

George, A., Sand provenance, in Antarctic Cenozoic history from the CIROS-1 drillhole, McMurdo Sound, *N. Z. DSIR Bull., 245,* 159–167, 1989.

Goldich, S. S., S. B. Treves, N. H. Suhr, and J. S. Stuckless, Geochemistry of the Cenozoic volcanic rocks of Ross Island and vicinity, Antarctica, *J. Geol., 83,* 415–435, 1975.

González-Ferrán, O., The Antarctic Cenozoic volcanic provinces and their implications in plate tectonic processes, in *Antarctic Geoscience,* edited by C. Craddock, pp. 687–694, University of Wisconsin Press, Madison, 1982.

Groushinsky, N. P., and N. B. Sazhina, Some features of Antarctic crustal structure, in *Antarctic Geoscience,* edited by C. Craddock, pp. 907–911, University of Wisconsin Press, Madison, 1982.

Hamilton, W., and E. L. Boudette, Quaternary alkaline basalt-trachyte province, Marie Byrd Land and south Victoria Land, Antarctica, paper presented at the International Symposium on Volcanology, Tokyo, May 9–19, 1962.

Harwood, D. M., Diatom biostratigraphy and paleoecology with a Cenozoic history of Antarctic ice sheets, Ph.D. dissertation, 590 pp., Ohio State Univ., Columbus, 1986.

Haugland, K., Seismic reconnaissance survey in the Weddell Sea, in *Antarctic Geoscience,* edited by C. Craddock, pp. 405–413, University of Wisconsin Press, Madison, 1982.

Hayes, D. E., and F. J. Davey, A geophysical study of the Ross Sea, Antarctica, *Initial Rep. Deep Sea Drill. Proj., 28,* 887–907, 1975.

Hayes, D. E., and L. A. Frakes, General synthesis, Deep Sea Drilling Project Leg 28, *Initial Rep. Deep Sea Drill. Proj., 28,* 919–941, 1975.

Hinz, K., and M. Block, Results of geophysical investigations in the Weddell Sea and in the Ross Sea, Antarctica, *Proc. World Pet. Congr. 11th, 1983,* 1–13, 1983.

Hinz, K., and Y. Kristofferson, Antarctica—Recent advances in the understanding of the continental shelf, *Geol. Jahrb. Reihe E, 37,* 1–54, 1987.

Höfle, H. C., The glacial history of the Outback Nunataks area in western north Victoria Land, *Geol. Jahrb. Reihe E, 38,* 335–355, 1989.

Hole, M. J., Post-subduction alkaline volcanism along the Antarctic Peninsula, *J. Geol. Soc. London, 145,* 985–998, 1988.

Hole, M. J., and W. E. LeMasurier, Geochemical variations and tectonic affinities of Cenozoic alkali basaltic rocks from the Pacific margin of West Antarctica (abstract), *Eos Trans. AGU, 71,* 1699, 1990.

Ishman, S. E., and P. N. Webb, Late Neogene foraminifera from the Victoria Land basin margin, Antarctica: Application to glacio-eustatic and tectonic events, *Rev. Paleobiol., Spec. Vol. 2,* 523–551, 1988.

Jankowski, E. J., Airborne geophysical investigations of subglacial structure of West Antarctica, Ph.D. dissertation, 293 pp., Cambridge Univ., Cambridge, England, 1981.

Jankowski, E. J., D. J. Drewry, and J. C. Behrendt, Magnetic studies of upper crustal structure in West Antarctica and the boundary with East Antarctica, in *Antarctic Earth Science,* edited by R. L. Oliver, P. R. James, and J. B. Jago, pp. 197–203, Australian Academy of Science, Canberra, 1983.

Johnson, B. D., C. M. Powell, and J. J. Veevers, Spreading history of the eastern Indian Ocean and greater India's northward flight from Antarctica and Australia, *Geol. Soc. Am. Bull., 87,* 1560–1566, 1976.

Johnston, A. C., Suppression of earthquakes by large continental ice sheets, *Nature, 330,* 467–469, 1987.

Kaminuma, K., Seismicity in Antarctica, in *Antarctic Geoscience,* edited by C. Craddock, pp. 919–923, University of Wisconsin Press, Madison, 1982.

Karl, H. A., E. Reimnitz, and B. D. Edwards, Extent and nature of Ross Sea unconformity in the western Ross Sea, Antarctica, in *The Antarctic Continental Margin: Geology and Geophysics of the Western Ross Sea, Earth Sci. Ser.,* vol. 5B, edited by A. K. Cooper and F. J. Davey, pp. 77–92, Circum-Pacific Council for Energy and Natural Resources, Houston, Tex., 1987.

Katz, H. R., Post Beacon tectonics in the region of Amundsen and Scott glaciers, Queen Maud Range, Transantarctic Mountains, in *Antarctic Geoscience,* edited by C. Craddock, pp. 827–834, University of Wisconsin Press, Madison, 1982.

Kellogg, K. S., and P. D. Rowley, Structural geology and tectonics of the Orville Coast region, southern Antarctic Peninsula, Antarctica, *U.S. Geol. Surv. Prof. Pap. 1498,* 25 pp., 1989.

Kim, Y., L. D. McGinnis, and R. H. Bowen, The Victoria Land basin: Part of an extended crustal complex between East and West Antarctica, in *Reflection Seismology: The Continental Crust, Geodyn. Ser.,* vol. 14, edited by M. Barazangi and L. Brown, pp. 323–330, AGU, Washington, D. C., 1986.

Klitgord, K. D., D. R. Hutchinson, and H. Schouten, U.S. Atlantic continental margin, Structural and tectonic framework, in *The Atlantic Continental Margin,* vol. I-2, *The Geology of North America,* edited by R. E. Sheridan and J. A. Grow, pp. 19–56, Geological Society of America, Boulder, Colo., 1988.

Kyle, P. R., Mineralogy and petrology of a basanite to phonolite sequence at Hut Point Peninsula, Antarctica, based on core from Dry Valley Drilling Project drillholes 1, 2, and 3, *J. Petrol., 22,* 451–500, 1981.

Kyle, P. R., McMurdo volcanic group—Western Ross Embayment, Introduction, in *Volcanoes of the Antarctic Plate and Southern Oceans, Antarct. Res. Ser.,* vol. 48, edited by W. E. LeMasurier and J. W. Thomson, pp. 18–25, AGU, Washington, D. C., 1990.

Lajoie, K. R., Coastal tectonics, in *Active Tectonics, Studies in Geophysics,* pp. 95–124, National Academy Press, Washington, D. C., 1986.

Lawver, L. A., and C. R. Scotese, A revised reconstruction of Gondwanaland, in *Gondwana Six: Structure, Tectonics, and Geophysics, Geophys. Monogr. Ser.,* vol. 40, edited by G. D. McKenzie, pp. 17–23, AGU, Washington, D. C., 1987.

Lawver, L. A., J. Y. Royer, D. T. Sandwell, and C. R. Scotese, Evolution of the Antarctic continental margin, in *Geological Evolution of Antarctica,* edited by M. R. A. Thomson, J. A. Crane, and J. W. Thomson, pp. 533–540, Cambridge University Press, New York, 1991.

LeBas, M. J., R. W. LeMaitre, A. Streckeisen, and B. Zanettin, A chemical classification of volcanic rocks based on the total alkali-silica diagram, *J. Petrol., 27,* 745–760, 1986.

LeMasurier, W. E., The Cenozoic West Antarctic rift system and its associated volcanic and structural features, *Geol. Soc. Am. Abstr. Programs, 10,* 443, 1978.

LeMasurier, W. E., Late Cenozoic volcanism on the Antarctic plate: An overview, in *Volcanoes of the Antarctic Plate and Southern Oceans, Antarct. Res. Ser.,* vol. 48, edited by W. E. LeMasurier and J. W. Thomson, pp. 1–17, AGU, Washington, D. C., 1990*a*.

LeMasurier, W. E., Marie Byrd Land, Summary, in *Volcanoes of the Antarctic Plate and Southern Oceans, Antarct. Res. Ser.,* vol. 48, edited by W. E. LeMasurier and J. W. Thomson, pp. 147–163, AGU, Washington, D.C., 1990*b*.

LeMasurier, W. E., and D. C. Rex, Volcanic record of Cenozoic glacial history in Marie Byrd Land and western Ellsworth Land: Revised chronology and evaluation of tectonic factors, in *Antarctic Geoscience,* edited by C. Craddock, pp. 725–734, University of Wisconsin Press, Madison, 1982.

LeMasurier, W. E., and D. C. Rex, Rates of uplift and the scale of ice level in stabilities recorded by volcanic rocks in Marie Byrd Land, West Antarctica, in *Antarctic Earth Science,* edited by R. L. Oliver, P. R. James, and J. B. Jago, pp. 663–670, Australian Academy of Science, Canberra, 1983.

LeMasurier, W. E., and D. C. Rex, Evolution of linear volcanic ranges in Marie Byrd Land, West Antarctica, *J. Geophys. Res., 94,* 7223–7236, 1989.

LeMasurier, W. E., and J. W. Thomson (Eds.), *Volcanoes of the Antarctic Plate and Southern Oceans, Antarct. Res. Ser.*, vol. 48, 487 pp., AGU, Washington, D. C., 1990.

Maslanyj, M. P., and B. C. Storey, Regional aeromagnetic anomalies in Ellsworth Land: Crustal structure and Mesozoic microplate boundaries within West Antarctica, *Tectonics, 9,* 1515–1532, 1990.

McGinnis, L. D., R. H. Bowen, J. M. Erickson, B. J. Allred, and J. L. Kreamer, East-west Antarctic boundary in McMurdo Sound, *Tectonophysics, 114,* 341–356, 1985.

McKelvey, B. C., P. N. Webb, D. M. Harwood, and M. C. G. Mabin, The Dominion Range Sirius Group—A record of the late Pliocene–early Pleistocene Beardmore glacier, in *Geological Evolution of Antarctica,* edited by M. R. A. Thomson, J. A. Crane, and J. W. Thomson, pp. 675–682, Cambridge University Press, New York, 1991.

Mercer, J. H., The Antarctic ice sheet during the late Neogene, in *Paleoecology of Africa and the Surrounding Islands,* edited by J. A. Coetzee, pp. 21–33, A. A. Balkema, Rotterdam, 1987.

Mueller, P., M. Schmidt-Thomé, H. Kreuzer, F. Tessensohn, and U. Vetter, Cenozoic peralkaline magmatism at the western margin of the Ross Sea, Antarctica, *Mem. Soc. Geol. Ital.*, in press, 1990.

Muir-Wood, R., Extraordinary deglaciation reverse faulting in northern Fennoscandia, in *Earthquakes at North-Atlantic Passive Margins, Neotectonics and Postglacial Rebound,* edited by S. Gregersen and P. W. Basham, pp. 141–173, Kluwer Academic, Boston, Mass., 1989.

Naeser, C. W., B. Bryant, M. D. Crittenden, Jr., and M. L. Sorenson, Fission-track ages of apatite in the Wasatch Mountains, Utah: An uplift study, *Mem. Geol. Soc. Am., 157,* 29–36, 1983.

Norton, I. O., and J. G. Sclater, A model for the evolution of the Indian Ocean and the breakup of Gondwanaland, *J. Geophys. Res., 84,* 6803–6830, 1979.

O'Connell, D. R. H., R. R. B. Von Frese, J. Paskievitch, and D. Damaske, Transantarctic Mountain (TAM) structure near Terra Nova Bay (TNB) from ship-to-shore seismic refraction profiles (abstract), *Eos Trans. AGU, 70,* 1220, 1989.

Okal, E. A., Intraplate seismicity of Antarctica and tectonic implications, *Earth Planet. Sci. Lett., 52,* 397–409, 1981.

Ostenso, N. A., and E. C. Thiel, Aeromagnetic reconnaissance of Antarctica between Byrd and Wilkes stations, *Res. Rep. Ser. 64-6,* p. 33, Geophys. and Polar Res. Cent., Univ. of Wis., Madison, 1964.

Parry, W. T., and R. L. Bruhn, Fluid inclusion evidence for minimum 11 km vertical offset on the Wasatch fault, Utah, *Geology, 15,* 67–70, 1987.

Pederson, D. R., G. E. Montgomery, L. D. McGinnis, C. P. Ervin, and H. K. Wong, Aeromagnetic survey of Ross Island, McMurdo Sound, and the dry valleys, in *Dry Valley Drilling Project, Antarct. Res. Ser.*, vol. 33, edited by L. D. McGinnis, pp. 7–25, AGU, Washington, D. C., 1981.

Prentice, M. L., G. H. Denton, T. V. Lowell, and H. C. Conway, Pre-late Quaternary glaciation of the Beardmore Glacier region, Antarctica, *Antarct. J. U.S., 21,* 95–98, 1986.

Robertson, J. D., C. R. Bentley, J. W. Clough, and L. L. Greischer, Sea-bottom topography and crustal structure below the Ross Ice Shelf, Antarctica, in *Antarctic Geoscience,* edited by C. Craddock, pp. 1083–1090, University of Wisconsin Press, Madison, 1982.

Robinson, E. S., Geological structure of the Transantarctic Mountains and adjacent ice covered areas, Antarctica, Ph.D. dissertation, 291 pp., Univ. of Wis., Madison, 1964*a*.

Robinson, E. S., Some aspects of subglacial geology and glacial mechanics between the South Pole and the Horlick Mountains, *Res. Rep. Ser. 64-7,* p. 88, Geophys. and Polar Res. Cent., Univ. of Wis., Madison, 1964*b*.

Robinson, E. S., and J. F. Splettstoesser, Structure of the Transantarctic Mountains determined from geophysical surveys, in *Geology of the Central Transantarctic Mountains, Antarct. Res. Ser.*, vol. 36, edited by M. D. Turner and J. F. Splettstoesser, pp. 119–162, AGU, Washington, D. C., 1984.

Roland, N. W., and F. Tessensohn, Rennick faulting—An early phase of Ross Sea rifting, *Geol. Jahrb. Reihe B, 66,* 203–229, 1987.

Roland, N. W., M. Olesch, and W. Schubert, Geology of the western border of the Transantarctic Mountains between Outback Nunataks and Reeves Glacier, map, *Geol. Jahrb. Reihe E, 38,* 1989.

Rooney, S. T., D. D. Blankenship, and C. R. Bentley, Seismic refraction measurements of crustal structure in West Antarctica, in *Gondwana Six: Structure, Tectonics, and Geophysics, Geophys. Monogr. Ser.*, vol. 40, edited by G. D. McKenzie, pp. 1–7, AGU, Washington, D. C., 1987.

Rowe, C. A., and J. Kienle, Seismicity in the vicinity of Ross Island, Antarctica, *J. Geodyn., 6,* 375–385, 1986.

Rowley, P. D., J. W. Thomson, J. L. Smellie, T. S. Laudon, K. E. La Prade, and W. E. LeMasurier, Alexander Island, Palmer Land, and Ellsworth Land, Summary, in *Volcanoes of the Antarctic Plate and Southern Oceans, Antarct. Res. Ser.,* vol. 48, edited by W. E. LeMasurier and J. L. Thomson, pp. 257–265, AGU, Washington, D. C., 1990.

Ruddiman, W. F., W. L. Prell, and M. E. Raymo, Late Cenozoic uplift in southern Asia and the American West: Rationale for general circulation modeling experiments, *J. Geophys. Res., 94,* 18,379–18,391, 1989.

Sandwell, D. T., and D. C. McAdoo, Marine gravity of the southern ocean and Antarctic margin from Geosat, *J. Geophys Res., 93,* 10,389–10,394, 1988.

Sato, S., N. Asakura, T. Saki, N. Oikawa, and Y. Kaneda, Preliminary results of geological and geophysical surveys in the Ross Sea and in the Dumont D'Urville Sea, off Antarctica, *Mem. Natl. Inst. Polar Res. Spec. Issue Jpn., 33,* 66–92, 1984.

Savage, M. L., and P. F. Ciesielski, A revised history of glacial sedimentation in the Ross Sea region, in *Antarctic Earth Science,* edited by R. L. Oliver, P. R. James, and J. B. Jago, pp. 555–559, Australian Academy of Science, Canberra, 1983.

Schmidt, D. L., and P. D. Rowley, Continental rifting and transform faulting along the Jurassic Transantarctic rift, Antarctica, *Tectonics, 5,* 279–291, 1986.

Schmidt-Thomé, M., P. Mueller, and F. Tessensohn, Malta Plateau, in *Volcanoes of the Antarctic Plate and Southern Oceans, Antarct. Res. Ser.,* vol. 48, edited by W. E. Masurier and J. W. Thomson, pp. 53–59, AGU, Washington, D. C., 1990.

Simpson, R. W., R. C. Jachens, R. J. Blakely, and R. W. Saltus, A new isostatic residual gravity map of the conterminous United States with a discussion on the significance of isostatic residual anomalies, *J. Geophys. Res., 91,* 8348–8372, 1986.

Smith, A. G., and D. J. Drewry, Delayed phase change due to hot asthenosphere causes Transantarctic uplift, *Nature, 309,* 536–538, 1984.

Smithson, S. B., Gravity interpretations in the Transantarctic Mountains near McMurdo Sound, Antarctica, *Geol. Soc. Am. Bull., 83,* 3437–3442, 1972.

Stern, T. A., and U. S. ten Brink, Flexural uplift of the Transantarctic Mountains, *J. Geophys. Res., 94,* 10,315–10,330, 1989.

Stock, J. M., Cenozoic plate motions in southwest Pacific region, paper presented at 28th International Geological Congress, Washington, D. C., July 9–19, 1989*a*.

Stock, J. M., Regional plate reconstructions of the New Zealand region since 68 Ma, paper presented at 28th International Geological Congress, Washington, D. C., July 9–19, 1989*b*.

Storey, B. C., and P. A. R. Nell, Role of strike slip faulting in the tectonic evolution of the Antarctic Peninsula, *J. Geol. Soc. London, 145,* 333–337, 1988.

Storey, B. C., et al., West Antarctic mosaic tectonic evolution, paper presented at 28th International Geological Congress, Washington, D. C., July 9–19, 1989.

Stuckless, J. S., and R. L. Ericksen, Strontium isotopic geochemistry of the volcanic rocks and associated megacrysts and inclusions from Ross Island and vicinity, Antarctica, *Contrib. Mineral Petrol. 58,* 111–126, 1976.

Sylvester, A. G., and S. W. Bies, Geodetic monitoring of fault movements in Death Valley, 1970–1985, in *Quaternary Tectonics of Southern Death Valley, California, Field Trip Guide,* edited by B. W. Troxell, pp. 41–44, B. W. Troxell (private publication), 1986.

Tessensohn, F., Passiv "Gondwanische" und aktiv "Pazifice" Tektonik der Antarktischen Platte, *Clausthaler Geol. Abh., Schoenberg Festschrift, 30,* 164–191, 1979.

Tessensohn, F., and G. Wörner, The Ross Sea rift system (Antarctica)—Structure, evolution, and analogues, in *Geological Evolution of Antarctica,* edited by M. R. A. Thomson, J. A. Crane, and J. W. Thomson, pp. 273–278, Cambridge University Press, New York, 1991.

Tingey, R. J., Uplift in Antarctica, *Z. Geomorphol. Suppl., 57,* 85–99, 1985.

Trehu, A. M., T. Holt, J. C. Behrendt, and J. C. Fritsch, Crustal structure in the Ross Sea Antarctica: Preliminary results from GANOVEX V (abstract), *Eos Trans. AGU, 70,* 1344, 1989.

Truswell, E. M., Geological implications of recycled palynomorphs in continental shelf sediments around Antarctica, in *Antarctic Earth Science,* edited by R. L. Oliver, P. R. James, and J. B. Jago, pp. 394–399, Australian Academy of Science, Canberra, 1983.

Webb, P. N., Late Mesozoic–Cenozoic geology of the Ross sector, Antarctica, *J. R. Soc. N. Z., 11,* 439–446, 1981.

Webb, P. N., The Cenozoic history of Antarctica and its global impact, *Antarct. Sci., 2,* 3–21, 1990.

Webb, P. N., and D. M. Harwood, Terrestrial flora of the Sirius formation: Its significance for late Cenozoic glacial history, *Antarct. J. U.S., 22,* 7–11, 1987.

Webb, P. N., and J. H. Wrenn, Upper Cenozoic micropaleontology and biostratigraphy of eastern Taylor Valley, Antarctica, in *Antarctic Geoscience,* edited by C. Craddock, pp. 117–1122, University of Wisconsin Press, Madison, 1982.

Webb, P. N., D. M. Harwood, B. C. McKelvey, J. H. Mercer, and L. D. Stott, Cenozoic marine sedimentation and ice-volume variation on the East Antarctic craton, *Geology, 12,* 287–291, 1984.

White, P., Downhole logging, in Antarctic Cenozoic history from the CIROS-1 drillhole, McMurdo Sound, *N. Z. DSIR Bull., 245,* 7–14, 1989.

White, R., and D. McKenzie, Magmatism at rift zones: The generation of volcanic continental margins and flood basalts, *J. Geophys. Res., 94,* 7685–7729, 1989.

Woollard, G. P., Crustal structure in Antarctica, in *Antarctic Research, The Matthew Fontaine Maury Memorial Symposium, Geophys. Monogr. Ser.,* vol. 7, edited by H. Wexler, M. J. Rubin, and J. E. Caskey, Jr., pp. 53–73, AGU, Washington, D. C., 1962.

Wörner, G., and L. Viereck, The Mt. Melbourne volcanic field (Victoria Land, Antarctica), I, Field observations, *Geol. Jahrb. Reihe E, 38,* 369–393, 1989.

Wörner, G., L. Viereck, J. Hertogen, and H. Niephaus, The Mt. Melbourne volcanic field (Victoria Land, Antarctica), II, Geochemistry and magma genesis, *Geol. Jahrb. Reihe E, 38,* 395–433, 1989.

Wrenn, J. H., and P. N. Webb, Physiographic analysis and interpretation of the Ferrar Glacier–Victoria Valley area, Antarctica, in *Antarctic Geoscience,* edited by C. Craddock, pp. 1091–1099, University of Wisconsin Press, Madison, 1982.

Zoback, M. L., et al., Global patterns of tectonic stress, *Nature, 341,* 291–298, 1989.

(Received August 22, 1990;
accepted April 8, 1991)